AF521381

Plant Physiology
in Relation to Horticulture

Second Edition

Plant Physiology in Relation to Horticulture

Second Edition

J.K.A. Bleasdale

Director
National Vegetable Research Station
Wellesbourne, Warwickshire

United Book Prints

First Reprint in India
United Book Prints
4806/24, Ansari Road, Daryaganj
New Delhi - 110 002
Tel.: +91-11-41511055
E-mail: unitedbookprints@gmail.com

First Indian Reprint 2026

ISBN: 978-81-926860-9-7

© J.K.A. Bleasdale 1973, 1984

CONTENTS

PREFACE

This book is primarily intended for students of horticulture who are studying for a Higher Diploma or a National Diploma in Horticulture, or for a university degree in which plant physiology is not a major option. However, as it describes the science behind modern horticulture, I hope anyone who is interested in gardening or plants will enjoy reading it – a deeper understanding of the principles behind horticultural practices should be enriching.

Graduates will also find this book useful if plant physiology enters into their work. For example, the plant breeder who needs to control flowering, produce good seed and compare the growth of breeding lines will find much of relevance here. Similarly, entomologists, mycologists, virologists and chemists who deal with plants should understand their behaviour better after reading this book and hence be more able to discuss their problems with plant physiologists.

Where data or extensive conclusions are quoted then the reference to the original work is given, and some suggestions for further reading are also listed. This limited acknowledgement of sources is intended to make for smoother reading, but it does not detract from my great indebtedness to those authors whose works were consulted, even though they are not specifically cited here.

In revising the book I was surprised how little needed to be changed, even after an interval of ten years. Perhaps I should not have been, since we are dealing here with basic processes in a basic manner. Nevertheless, the chapter on 'The Seed' has been considerably amended and the other chapters have been updated to take account of the new work published in the last decade. I am grateful to several colleagues for assisting me in the task of deciding the extent and nature of the revisions.

I am also grateful to those colleagues, authors and publishers who have allowed me to use their data or diagrams.

1983 J. K. A. BLEASDALE

CHAPTER 1

THE SEED

Many horticulturists must have been introduced to their subject by becoming enchanted by the picture on some seed packet. These simple packets may be all that is necessary to protect the seed during its months or even years in conditions unsuitable for growth. This remarkable property of self-preservation in a wide range of conditions, this ability to wait until conditions become favourable for growth, is thought of as the major Darwinian advantage leading to the success of flowering plants. Man has exploited this property of seed for trading plants across continents and oceans in centuries when rapid travel was only for the gods. The food reserves in seed are easily stored and often are an important food source for man in seasons or years when fresh produce is not available. The Bible tells of the corn that Joseph stored in Egypt against the seven lean years and the importance of grain crops for world food supply is still obvious.

The diversity of the form and content of seed is legion. Such extremes as the coconut (*Cocus nucifera*) and mustard (*Sinapis alba*) are known to everyone, but even dehydrated coffee comes from seed, and soap and margarine are usually made with oil derived from seed. Many drugs are also still derived from seed and trade in the raw materials of all these commodities is made easier because seed can tolerate, without marked deterioration, a wide range of conditions. Yet all that is needed to start most seed growing is the simple and abundant chemical – water.

The seed takes up water firstly by imbibition, during which the desiccated colloidal contents of the seed become soaked. Imbibition is accompanied by swelling of the seed and occurs regardless of whether or not the seed is alive, even in the absence of air and at low temperatures. This rehydration is essentially similar to what happens when water is added to dehydrated vegetables in packet soups. Great forces can develop within dry seed that is imbibing water – up to 200 Mpa; that is why it is dangerous to soak seed in a closed container, be it a screw-top jar or the leaky hull of a ship.

As more water enters the seed by imbibition the suction force diminishes but the hydrated membranes and protoplasm begin to allow *osmotic forces* to operate, facilitating further rehydration. The tissues regain the size and shape they had before they ripened, and as the fine structure of the cells is restored the metabolic activity (the chemistry of growth) begins. Air is essential for this metabolic activity to take its normal course. When there is little oxygen about, growth ceases and the seed enters a state of **induced dormancy**, remaining capable of regrowth.

When oxygen is completely lacking, the seed will die because the products of anaerobic metabolism are lethal. For example, considerable quantities of alcohol and lactic acid accumulate in imbibed pea seeds (*Pisum sativum*) in the absence of air.

Many rehydrated seeds will not grow even when abundantly supplied with air. These seeds, provided they can subsequently be shown to be viable, are said to be dormant. Dormancy has many causes and the horticulturist should be aware of some of the major ones and how to overcome them.

SEED DORMANCY

Impermeable seed coats can restrict the entry of water or gaseous exchange, so that germination is prevented even in the presence of abundant water and air at normal temperature. If some imbibition can occur, then the seed will slowly swell and sometimes burst a restrictive seed coat, so making further growth possible.

Hard seed coats will rot if buried in soil, and it is in this way that the impermeable coats on seeds of some common weeds are softened. For example, plants of Fat Hen (*Chenopodium album*) produce two types of seed — one with a normal testa and the other with a hard seed coat. This ensures that even after several years of clean cultivation some seed is almost ready for imbibition and growth. Other species, many of them belonging to the important legume family, also produce varying proportions of hard seed.

Hard seed can be easily detected by soaking a small sample in water for about 12 hours. If some seeds do not swell, then it is reasonable to suppose that an impermeable seed coat is restricting imbibition. Several treatments can render these seed coats permeable, but the simplest and safest is mechanical scarification. This treatment is safe because soft seeds are not likely to be damaged. Dry seed may be vigorously shaken so that the seeds abrade one another. With harder coats gentle rubbing between two sheets of sandpaper can be effective. These mechanical treatments have the further advantage that the seed remains dry and can therefore be easily sown by hand or machine.

Soaking seed in concentrated sulphuric acid can also be very effective if somewhat hazardous. Seed should be placed in about one and a half times its own volume of acid, stirred frequently but carefully, and then cautiously plunged into cold water and washed thoroughly. The treatment time may vary from a few minutes up to two hours depending on the type of seed. For example, the hard seed of *Cytisus scoparius* (Scotch Broom) needs to be soaked in acid for at least one hour but will tolerate up to five hours before germination is impaired. Seed treated with acid can be dried and sown later, but some loss of viability does occur after several months of dry storage.

Soaking in absolute ethanol for three days can also overcome the impermeability of some seed, notably those of the sub-family Caesalpinioideae of the family Leguminosae. Here the treatment is more effective if the seed is sown soon afterwards. This treatment is not more effective than mechanical scarification but its mode of action and selectivity, which are not fully understood, make it of passing interest.

Some seeds, notably those of trees, may need further treatment in addition to scarification before they will grow.

Stratification (the moist, low-temperature treatment of seed) overcomes the embryo dormancy of the seed of more than 100 species, mainly perennial plants and not uncommonly trees. The need for stratification ensures that the seed does not germinate until the passing of winter and the return of favourable growing conditions. Seed that passes through the winter in moist soil receives the necessary stimulus naturally. Several weed species require their imbibed seed to encounter a period of low temperature before they will germinate; one example of this, which will be discussed in some detail later, is found in *Polygonum aviculare*.

Horticulturally, seeds are stratified by burying them in a tray of moist sand and exposing them to a temperate winter, or by placing them in a refrigerator at about 5°C for 8 to 20 weeks. Germination will occur when seed treated in this way is warmed; this effect can be readily demonstrated by using freshly harvested apple or rose seed. *Cotoneaster divaricata* and *C. horizontalis* are examples of species needing scarification before stratification is effective.

Epicotyl dormancy is a relatively rare type of dormancy removed by low temperature. This type of dormancy is well exhibited by the Tree Paeony (*Paeonia suffruticosa*) and *Viburnum opulus*. Their seed germinate normally in that the radicle emerges from the testa and begins to grow, but the shoot (epicotyl) will not grow until the seed has been stratified. Three to six months at a high temperature (20°C) is needed for the root to grow from the seed of some species of *Lilium* but the shoot will not grow until the seed has been kept at a low temperature (below 1.0°C) for six weeks.

Light is required by the seed of some species for germination to follow rehydration. Much has been written about the mechanism involved, mainly about the lettuce cultivar Grand Rapids. It is well established that seed of this cultivar and several others are light-requiring when freshly harvested but it should not be assumed that all lettuce cultivars behave similarly. Furthermore there are several other relatively common horticultural species whose germination is promoted by light (for example, some cultivars of celery, tobacco and tomato).

In common with other light-induced developmental responses in plants (termed **photomorphogenic responses**), this light-sensitivity depends on the pigment **phytochrome** (see chapter 5). The red region of the spectrum (650 nm) promotes the germination of *positively photoblastic* seed, whereas far-red radiation (700 nm) inhibits their germination.

Very little light is often all that is required to stimulate germination. For example, only 100 lx s (lux-seconds) are needed to promote the germination of tobacco (*Nicotiana tabacum*) seed, and the brief exposure to light of some weed seed during tillage has been shown to promote their germination.

Darkness promotes the germination of *negatively photoblastic* seed and this is commonly achieved by covering seed with soil. *Nigella damescena* is the best-known example of a negatively photoblastic species. The pigment phytochrome is again involved in this response.

After-ripening, which can occur in dry storage, is needed by many seed after harvesting. Common examples of seed needing after-ripening are those of wheat, oats and barley. However, the seed of the aquatic plant *Rorippa nasturtium aquaticum* (the commercial watercress) have to be stored in a moist atmosphere to complete their ripening. Seed requiring stratification need special low temperatures for after-ripening. In contrast the seed of wheat, oats and barley can be after-ripened by two days storage at 35 to 40°C.

Freshly harvested lettuce seed also require after-ripening for they will not germinate at temperatures consistently above about 20°C, even when sown under otherwise ideal conditions. After only a few weeks' dry storage the seeds will germinate at 20°C but may remain dormant at higher temperatures. Keeping the moist seed at a low temperature (5°C) or soaking in a 1 per cent solution of thiourea will overcome this dormancy in the fresh seed.

Induced dormancy is the restoration of dormancy to seed that could previously grow at normal temperatures in the presence of water and air. Induced dormancy is significant in regulating the periodicity of weed seed germination in temperate climates. With the seed of both weeds and cultivated plants this secondary dormancy is frequently induced either by a period at high temperature (about 25°C), by a restricted oxygen supply or by a combination of these two factors.

Courtney (1968) showed that in the field the seed population of the weed *Polygonum aviculare* undergoes cyclic changes in physiological dormancy which govern seedling emergence. The seed freshly harvested from the plant were in a state of innate dormancy, which was overcome by exposure to low temperatures (2 to 4°C) while moist. Thus, freshly shed seed, having passed its first winter in the soil, was able to germinate as soon as the temperature began to rise. Some of these seed did not emerge – for reasons that are not precisely understood. However more seedlings emerged from cultivated soil than from undisturbed soil; Courtney concluded that either lack of a light stimulus or excessively high levels of carbon dioxide in the soil might be responsible for the dormancy. Such seed are able to germinate because they are not dormant but nevertheless do not germinate even though they are moist and at normal temperatures: they are said to be in a state of enforced dormancy. In this instance the dormancy was thought to be enforced by the lack of light or too much carbon dioxide; removal of this restriction resulted in germination.

When the soil temperature rises, those seed in enforced dormancy enter a state of induced dormancy. Courtney showed (see table 1.1) that as little as two weeks at 25°C induced a state of dormancy that permitted little germination at temperatures above 8°C, whereas about half the seed had previously been capable of doing so. Induced dormancy differs from enforced dormancy in that removal of the cause does not result in germination. It is similar to innate dormancy in that exposure to some specific stimulus is required to overcome the dormancy.

Table 1.1

Percentage germination of apparently viable seed of *Polygonum aviculare* after different periods at 25°C in undisturbed soil (after Courtney, 1968)

Weeks at 25°C in soil	*Temperature of germination (°C)*				*Days to 50 per cent germination at 4°C*
	4	8	12	23	
0	84.6	54.1	63.1	41.2	21
1	97.6	36.1	39.8	27.7	24
2	95.4	3.4	3.4	8.0	102
3	80.8	1.2	1.2	1.2	108
4	91.1	0.0	3.2	1.6	107

Induced dormancy develops most rapidly in the field in seed near the soil surface. This accords with the soil temperature profile, suggesting that temperature-induced dormancy is one of the factors contributing to the cyclic emergence of the seed in the field. The drying of *P. aviculare* seed leads to induced dormancy, possibly reinforcing the temperature effect and ensuring that this weed only emerges when it is likely to be able to complete its life-cycle.

Chemical inhibitors are implicated in many of the types of dormancy already discussed. The role of phytochrome in photoblastic seed can be considered in this context, as can the inhibitor which can be extracted from the many seed that have to be stratified. Stratification generally leads to the production of substances that promote germination rather than to the destruction of inhibitor. However, some inhibitors in dormant seed can be easily removed by leaching with water. Very often such seed are not true seed in the botanical sense but are seed enclosed in pericarp. For example, the commercial seed of red beet and sugar beet (*Beta* spp.) are fruits and their pericarp contains a water-soluble germination inhibitor.

Germination inhibitors are also present in many succulent fruits. It is supposed that such inhibitors play a part in ensuring that seed do not germinate inside the fruit, where conditions are seemingly ideal for germination.

Apparently the normal functioning of this inhibitory mechanism in fruits may depend on an adequate supply of potassium. When potassium was withheld from pepper plants growing in nutrient solution, Harrington (1960) found that more than 6 per cent of the seed germinated within the fruit, some fruits being a solid mass of radicles and hypocotyls when cut open. In control treatments sprouting seed were very rare and these only had very short radicles.

Germination of seed while still attached to the parent plant is termed **vivipary**. Because vivipary is considered a disadvantage by horticulturists, it is not found in cultivated plants unless they are subjected to abnormal conditions such as those just described. However, sprouting of cereal grains in the ear is common in some wheat varieties during wet summers and in some wild species it is highly developed to ensure survival. The seed of the Mangrove would be washed away by the tides if they did not germinate and root whilst still attached to their parent.

GERMINATION TESTS

> One of the great hazards in agriculture is sowing seed that has not the capacity to produce an abundant crop of the required cultivar. Seed testing has been developed to minimise this risk by assessing the quality of seed before it is sown. Seed quality is a concept made up of different attributes. These attributes are of interest to different segments of the industry — to the producer, the processor, the warehouseman, the merchant, the farmer, the certification authority and to the government or agency responsible for seed control. In all cases the ultimate object of making a test is to determine the value of seed for planting.

This laudable statement of the history and objectives of seed testing begins the introduction to the *International Rules for Seed Testing* (Anon. 1976). It emphasises that seed testing has objectives and its origin demonstrates that to achieve these the tests must be standardised and internationally agreed.

Seed testing covers other facets of seed quality besides those evident in germination tests. Freedom from rubbish or foreign seed — that is, purity testing — determination of moisture content, state of health and trueness of type are all facets of the seed analyst's profession. We are here concerned with the germination test designed to gain information with respect to the field planting value of seed and to provide results that can be used to compare the value of different seed lots.

The need for standard tests that can be reproduced at different centres by different operators and at any time of year – together with the need for rapid results – has led to higher temperatures being used for tests than are encountered by most crops in the field. Therefore, the germination test determines the proportion of the seed that is viable but in practice only some of these viable seed are able to emerge in the field. The analytical procedure emphasises that a seed must not be considered viable unless the "essential structures" are sufficiently developed to allow abnormality to be detected.

For many species the standard test procedure is designed to overcome any normal state of dormancy. If apparently viable seeds remain ungerminated after the prescribed test, the analyst may choose to apply a dormancy-breaking treatment to them and subsequently reassess their viability. This is always fully reported on the test certificate issued by recognised seed-testing laboratories.

Small seed are usually tested by laying them on moistened filter papers, whereas large seed such as peas or beans are laid between paper or in moistened sand. These substrates are moistened with either good quality tap water or deionised water. To overcome dormancy a 0.2 per cent solution of potassium nitrate is used for the initial wetting. For reasons that are not understood, nitrates will often overcome the dormancy associated with after-ripening or tend to replace a requirement for light. If too much water or nitrate solution is used, aeration is reduced and the germination adversely affected. As water is imbibed by the seeds or lost by evaporation, it should be replenished. The need for replenishment is minimised by containing the substrate in a Petri-dish or water is supplied by a wick as in the Copenhagen tank diagrammatically represented in figure 1.1.

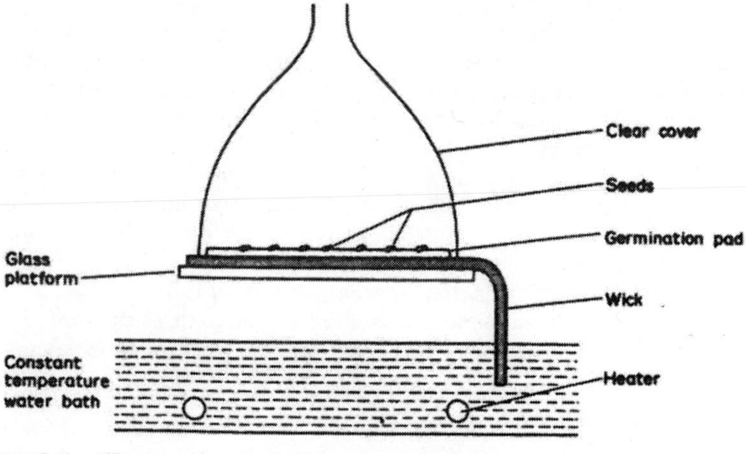

Figure 1.1 Diagram of one unit of a Copenhagen tank as used in germination tests.

As previously mentioned the standard temperatures are usually somewhat higher than those encountered in the field and may be lower for a 16 hour night period than for an 8 hour day period. If a germination test is to be carried out on a species of horticultural importance, it is usually most effective to use the procedure set out in the current *International Rules for Seed Testing*, the full reference to which is given at the end of this chapter.

However, the purpose of a test may preclude the adoption of the standard conditions. This is often the case when studying the effects of different chemical substances on seedling emergence, because sub-optimal conditions (particularly of temperature) are more likely to reveal the effects of the specific chemical treatment. Regardless of the nature of the test, results are obtained by counting (usually as they are removed) the seed which germinate and exhibit the potential to develop into normal plants. In the standard tests it is usual to make two counts at specified intervals, one early and the other at the end of the period of expected germination. If most of the seed germinate and the early count is a high proportion of the late count, then the seed have passed with flying colours.

Some seed, notably those of trees, germinate so slowly that a more rapid method of assessing their viability than the standard germination test is often required. Because dead cells will imbibe water, the swelling of seed is no indication of their viability. The rapid test must show that the biochemical processes of the seed are functioning. This can be revealed by the vital stain, tetrazolium chloride (or bromide). In living cells the essential reduction processes, in which hydrogen is accepted from dehydrogenases, are active. The colourless solution of the tetrazolium salt imbibed by the seed cells accepts hydrogen to form a stable and non-diffusible red substance, triphenyl formazan.

Those seed which stain completely red are viable, whereas those without any coloration are dead. However, there are usually some seed which are partially stained and the decision as to their viability depends on the nature of the living parts. Again the *International Rules for Seed Testing* should be consulted for guidance in making these decisions and for details of techniques.

Usually the test is made on seed that have been soaked in water for 18 to 20 hours. Typically the imbibed seed are then immersed in a 1 per cent aqueous solution of tetrazolium chloride (or bromide) for 24 hours in the dark at 30°C, after which they are rinsed and examined. Between soaking and immersion in the solution, various 'surgical' operations are performed, such as snipping off one-third to one-sixth of the seed. After staining it may be necessary to dissect out the embryos to assess which have stained.

RATE OF GERMINATION

The early and late counts made in many standard seed tests indicate the rate of germination. If the two counts are similar the rate of germination is high, and *vice-versa*.

In general the rate of seed germination rises with increasing temperature up to about 35°C (some exceptions associated with dormancy have already been described) and this is also true of emergence in the field. The rate of emergence is often conveniently expressed as a single figure, denoting the number of days before half the emerged seed appeared. Figure 1.2 illustrates the effect of soil temperature on rate of emergence.

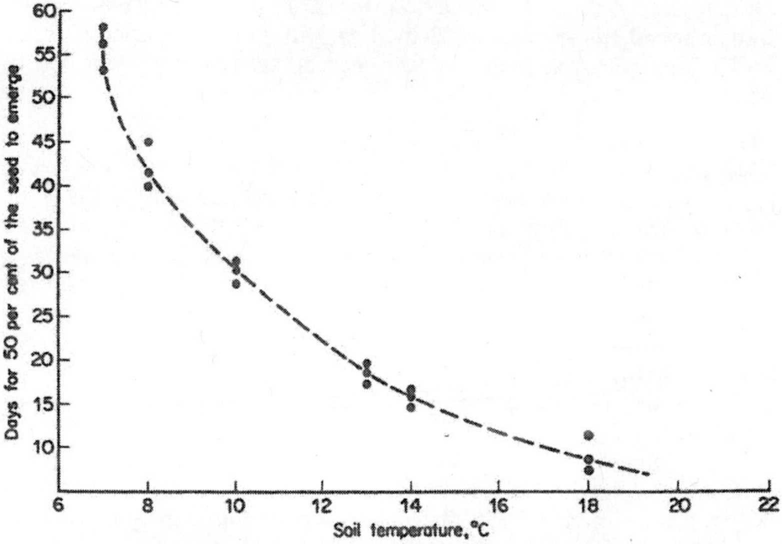

Figure 1.2 The effect of soil temperature on the days to 50 per cent emergence of seed of *Phaseolus vulgaris* cv. Glamis (data R. C. Hardwick, National Vegetable Research Station).

This relationship between temperature and rate of emergence is not characteristic of a genus nor of a species within a genus. Different cultivars of *Phaseolus vulgaris* can show different effects of temperature on their rates of emergence. Nevertheless, some species are generally slow to germinate and emerge whereas others are rapid.

We have already seen that germination can be regarded as the result of imbibition and growth. Imbibition was likened to the reconstituting of dehydrated vegetables. Hydration by soaking is the first stage in the canning of processed peas (the seed of pea plants). Processors noticed that if the initial moisture content of the dried peas was below 10 per cent, then the uptake of water was uneven, some peas remaining hard and unswollen while others were fully swollen (Crain and Haisman, 1963). This effect is also apparent if dry pea seed are sown in soil or subjected to a germination test. The variation in the rates of imbibition by seed leads to a slower rate of emergence or germination. Similar effects have also been reported with *Phaseolus* and it seems probable that this is a general phenomenon.

The variation in imbibition rates can be of considerable importance when, for example, seed from different sources are sown in a trial of cultivars. The differences in rates of emergence, due to variation in moisture content of seed, could still be apparent at harvest-time, giving a meaningless indication of true performance. Such ambiguity is best avoided by storing all the samples for 2 to 3 days in a saturated atmosphere at 25°C. This will ensure that all the seed are sufficiently moist before planting.

Rate of respiration.

During hydration the growth processes are revitalised. The soluble sugars in the seed are soon used up by respiration and the cells of the plant embryo begin to enlarge and divide. Enzymes are activated or synthesised and insoluble storage compounds are hydrolysed: proteins by proteinases and peptidases; lipids by lipases; starch by amylase, maltase or phosphorylases. The cells of the embyro continue to elongate and divide, and the positively geotropic radicle emerges (see chapter 3). The plumule grows upwards (negatively geotropic) and as it breaks the surface of the substrate, photosynthesis commences and the new plant becomes established.

The intense activity during germination is indicated by the high rates of respiration and the changes in the biochemical pathways as development continues. The changing pattern of this activity is reflected by changes in the **respiratory quotient** (RQ) which is simply the ratio of the volume of CO_2 evolved to the volume of oxygen consumed.

During aerobic oxidation the RQ characterises the chemical substrate being respired: carbohydrates 1.0; proteins, 0.8 to 1.0; and lipids 0.7 to 0.8. The ranges of the proteins and lipids reflect the diversity of different chemical structures. Early in germination, RQs higher than unity have been recorded and interpreted as indicative of oxygen starvation or the release of carbon dioxide trapped within the dry seed.

The rate of respiration of imbibed seed is a measure of biochemical activity and hence of the rate of germination. Figure 1.3 shows the rate of oxygen uptake by dormant and after-ripened seeds of *Avena fatua*. During the first few hours the rate increases in both kinds of seed but after about 16 hours the lower rate of the dormant seed no longer increases. At about this time, the non-dormant seed germinate (the radicle emerges) and the respiration rate continues to rise.

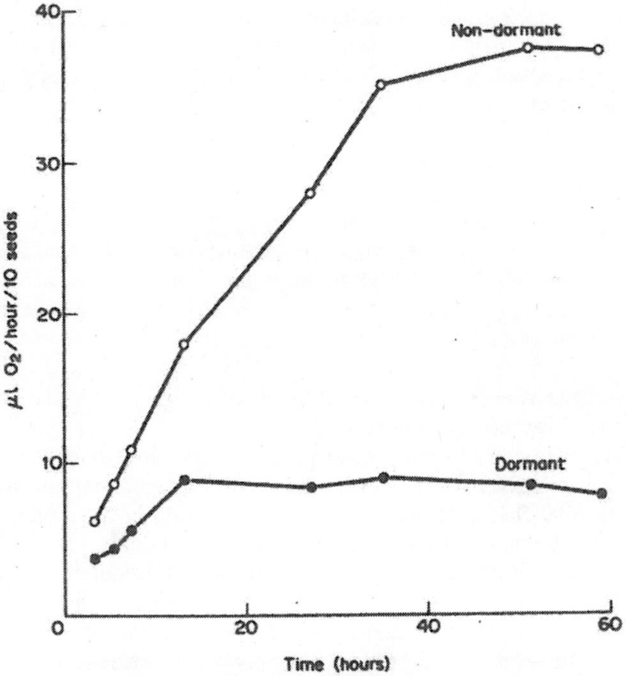

Figure 1.3 Time course of respiration of dormant and non-dormant seed of *Avena fatua* at 21.5°C (after Shapley and Varner, 1970).

Detailed studies of this kind have revealed more about the nature of dormancy. From the work on *Avena fatua* it was concluded that dormancy was not a state of general inactivity but a curtailment of a specific metabolic process.

Respiration rates rise with an increase in temperature up to levels where injury occurs (40°C). Within this range, respiration rates almost double for every 10°C rise. Low temperatures frequently occur in the field and in theory these should only delay emergence but in practice many of the seed fail to emerge if germination is too much protracted. Some seeds which are viable in a germination test do not produce plants in the field. These tests can only be of full practical value if interpreted correctly for field use. Otherwise they must be modified to make them more valuable.

LABORATORY GERMINATION AND FIELD EMERGENCE

The early and late counts made in standard germination tests give some indication of the rate of germination. Several workers found that a high germination rate was indicative of a good performance in the field but this has not provided the basis for predictive tests because it has proved difficult to standardise procedures for measuring germination rates. In addition, other methods of measuring vigour are more repeatable and are now accepted as being of considerable value in predicting the field performance of seed.

The International Seed Testing Association Congress in 1977 adopted the following definition of seed vigour:

> Seed vigour is the sum total of those properties of the seed which determine the level of activity and performance of the seed or seed lot during germination and seedling emergence. Seeds which perform well are termed high vigour seeds and those which perform poorly are called low vigour seeds.

Over recent years several new tests for seed vigour have been proposed and some older tests have been critically examined (Perry, 1981). One of the newer tests stems from the observations of several workers that samples of pea (*Pisum sativum*) seed with similar laboratory germinations often gave very dissimilar field emergences and that this could be attributed to some lots of seed being predisposed to attack by the soil-borne fungus, *Pythium ultimum*. Dressing the seed with a fungicide, a frequent horticultural practice, increases field emergence somewhat, but even when this is done it is not uncommon that only half the viable seed emerges to form plants.

In a study of some sixty-three lots of pea seed, Matthews and Bradnock (1967) found that the relationship between laboratory germination and field emergence was not significant (correlation coefficient 0.20 to 0.48). However, if they soaked twenty seed in 200 ml of deionised water for 24 hours at 15°C, there was a good relationship between the conductivity of the soak water and field emergence (correlation coefficient of −0.6 to −0.85). Seed lots which exuded chemical substances into the soaking water, thereby increasing their conductivity, gave poor field emergence even though their laboratory germination was high and they were dressed with a fungicide.

Matthews has subsequently presented evidence suggesting that exudation is typical of groups of cells prone to invasion by the fungus. These cells may be in some way damaged when the seed is being threshed and the damage may be particularly extensive if the seed crop is immature. Danish workers (Jensen and Jørgensen, 1969) have shown that immature seed of *Festuca pratensis* is more prone to threshing damage than mature seed with a similar moisture content.

The response of seed to accelerated ageing is also proving to be of value in assessing their vigour. The seeds under test are subjected to periods at high temperature and high relative humidity. The precise temperature varies with the species under test. For onion (*Allium cepa*) 72 hours at 42°C is suggested, for radish (*Raphanus sativus*) 48 hours at 45°C, and for water-melon (*Citrullus lanatus*) 72 hours at 45°C. These ageing treatments reduce the viability of the seed as measured in the standard germination test. If the reduction is very marked the seed is likely to be of low vigour and *vice versa*.

It seems probable that the best predictions of field performance will come as a result of interpreting the standard germination test in the light of several different vigour tests. There is no doubt that as our ability to predict seed bed conditions is limited it is invariably safer to start with seed of high vigour. The practical horticulturist is, however, faced with deciding on a sowing rate using the seed available.

Calculation of sowing rates

Horticulturists are increasingly concerned that they should sow at rates which will achieve some required population without recourse to expensive thinning by hand. I have suggested that growers should calculate the sowing rate by using the equation

$$\text{Seed required in grammes per hectare}^{\dagger} = \frac{\text{mean weight per seed in mg} \times \text{number of plants required per m}^2}{\text{per cent laboratory germination} \times \text{field factor}}$$

The field factor (FF) in this equation takes account of the common experience that field emergence is lower than laboratory germination. An FF of unity would mean all the viable seed were expected to emerge. In practice, under good conditions one might use FF = 0.8 and under poor conditions a value of 0.4. In practice the equation has been useful because it has enabled growers to avoid the errors which can arise if similar weights per unit area are sown each year without regard to seed size, percentage laboratory germination or field conditions.

The relationship between the laboratory and field performances of seed that the equation assumes is shown in figure 1.4a. Some sets of data suggest, at least for some crops, that it may be as effective or even better to subtract a constant from the laboratory germination and leave out the FF term. This has the same effect as using a higher FF if the laboratory germination is low and is illustrated in figure 1.4b. The dotted line in this figure shows the effect of an FF = 0.8; it would clearly be difficult to distinguish between these two models over the limited range of laboratory germination encountered in commercial seed samples.

However, it seems likely that the results of vigour tests should be taken into account in assessing the FF term. Under a wide range of conditions seed of high vigour would be expected to give populations in line with

† Alternatively

$$\text{Seed required in lb per acre} = \frac{272 \times \text{number of plants required per sq.ft.}}{\text{number of seeds per oz in thousands} \times \text{per cent laboratory germination} \times \text{field factor}}$$

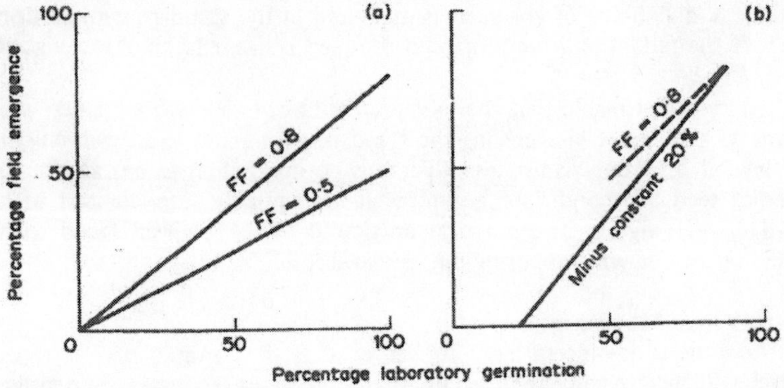

Figure 1.4. The assumed relationship between the laboratory germination and the field emergence of seed. (a) Using two values of the field factor (FF) and (b) an FF = 0.8 compared with subtracting 20 from the percentage laboratory germination.

those predicted by the laboratory germination test. Such seed would nearly always warrant the use of a high FF value. On the other hand, seed of low vigour would perform relatively well under good conditions, perhaps even warranting as high an FF as vigorous seed, but under poor conditions the appropriate FF value would be much lower. The prudent grower will make counts of his actual population for each sowing, so that he may learn by his mistakes and successes.

Another approach to achieving some required population by direct-seeding without thinning is to select from a normal seed sample those seed that are most likely to emerge in the field. In addition, these seed may be treated in some way to enhance further their chances of growing into plants. This approach is particularly relevant where regularly and widely spaced plants are required such as with lettuce or cabbage.

SEED TREATMENTS GIVING ENHANCED PERFORMANCE

Seed size

In general, the larger and heavier seed in a sample are more likely to be viable. Thus, if the small seed are eliminated by sieving or winnowing, the laboratory germination and field emergence of the remaining sample is higher than that of the original seed population. Seedsmen 'dress' seed lots in this way to ensure that they are of good viability or above some statutory minimum level of viability.

Although larger seed usually have a higher laboratory germination than smaller ones, the largest seed do not necessarily give the highest percentage field emergence. This is illustrated by the data in table 1.2. The larger cotyledons of the larger seed seem to impede emergence. Some of the seedlings that do emerge may consequently lose one or both cotyledons, making them poor plants. This is only likely to occur in some epigeal dicotyledonous species.

Table 1.2

The effect of seed size on the germination and emergence of summer cauliflower cv. Meteor (after Hewston, 1964)

Diameter mm	*Mean weight per seed mg*	*Percentage laboratory germination*	*Percentage field emergence*
1.75–2.0	2.9	72	68
2.0 –2.5	4.1	81	82
2.5 –3.0	5.3	88	63

Vernalisation

We have already seen that cold can overcome the dormancy of some seeds. Low temperatures encountered by imbibed seeds can also promote flowering of some species, particularly winter annuals. This effect was first exploited by Lysenko, a Soviet geneticist of the Stalin era, who thought that the genetic make-up of the plants was altered by the low temperatures. We now know that this is not so, but many writers use the term derived from the Russian —*jarovisation* — to refer to the process. The more usual term (of latin derivation) is *vernalisation* which can be translated as 'belonging to spring'.

Vernalisation is the promotion of flower initiation by a previous cold treatment. As we shall see later in chapter 5, in most species that can be vernalised the seed must have developed into young plants before they respond to cold. However, Lysenko's exploitation of vernalisation and much of the earlier work on the physiology of the process were concerned with the conversion of typically 'winter' strains of cereals into 'spring' forms by the cold treatment of imbibed seed.

Starting in the late 1930s Gregory, Purvis and their colleagues at Imperial College published a series of papers on the vernalisation of the winter strain of Petkus Rye (*Secale cereale*), which appears to differ from the spring strain by a single gene only. The winter strain could be made to behave like the spring strain by subjecting imbibed seed to periods of low temperature. The similarity of the response of the two strains depended on the duration of the cold treatment. A few days of cold treatment had some effect but 4 to 6 weeks of cold were needed to produce nearly identical responses from the two strains. The effectiveness of the treatment was therefore quantitative — the longer the cold treatment the greater the effect. The nature of this quantitative relationship is complex. For short treatments of 10 days higher temperatures (10°C) are usually more effective than lower temperatures (1°C). For lower temperatures to become the more effective, the treatment period has to be increased. This observation has been interpreted as supporting the finding that vernalisation requires energy and needs oxygen and a chemical substrate, such as sugar.

Vernalised seed or plants can be devernalised by periods at high temperatures (20 to 35°C). If the period at low temperature is immediately followed by a period at an intermediate temperature, vernalisation is stabilised and subsequent high temperatures have little or no effect. Devernalisation is most likely to occur if the period of cold is immediately followed by high temperatures. Some devernalisation of Petkus Rye seed can occur when the seed is dry.

The seed of mustard (*Sinapis alba*) can be devernalised by drying if the testas split during vernalisation. Vernalised seed with intact testas have been stored at room temperature for 6 years without any significant devernalisation.

Several workers have reported that cold treatment of imbibed seed of some lettuce cultivars leads to an earlier production of flower stalks. Effective treatments ranged from 10 days at −5°C to 16 days at 8°C. Whether or not such a treatment would be regarded as promoting the seed performance would clearly depend on whether one was a grower of seed or head crops!

Peas (*Pisum sativum*) flower earlier with fewer nodes below the first flower if the seed are vernalised. Again, this is a *quantitative response* — flowering does not depend on a period at low temperature but is accelerated by such treatment. We will see later in chapter 5 that cold is essential for flowering of some plants and the response is then *qualitative*.

Germination advancement

Various seed treatments have become established which have in common the advancement of germination prior to sowing. One way of achieving such advancement is to subject the seed to short wetting and drying cycles. Following earlier Russian claims, Austin *et al.* (1969) working in England found that the best effects on carrot were obtained by wetting the seed with water equal to 70 per cent of its own weight, incubating the wetted seed for 24 hours at 20°C in covered containers, drying slowly over 2 days, and repeating the cycle twice more. This process has been termed *drought hardening* or *hydroisation*.

These seed had the same total germination as control seed but the time for half the viable seed to germinate was 3 to 4 days less than for untreated seed. Treated seed imbibed water at a much higher rate than untreated ones (figure 1.5), and during treatment the seed embryos enlarged mainly as a result of cell division. In a field experiment using five different dates of sowing, 3 to 8 week old seedlings from treated seed had dry weights about 50 per cent greater than those from untreated seed. Harvests made at 15 and 19 weeks after sowing gave an average yield increase of about 10 per cent. Fluctuations in the increases could not be related to the dryness of the soil at sowing. Austin *et al.* (1969) concluded that the increased yield could be accounted for by earlier emergence and greater embryo size.

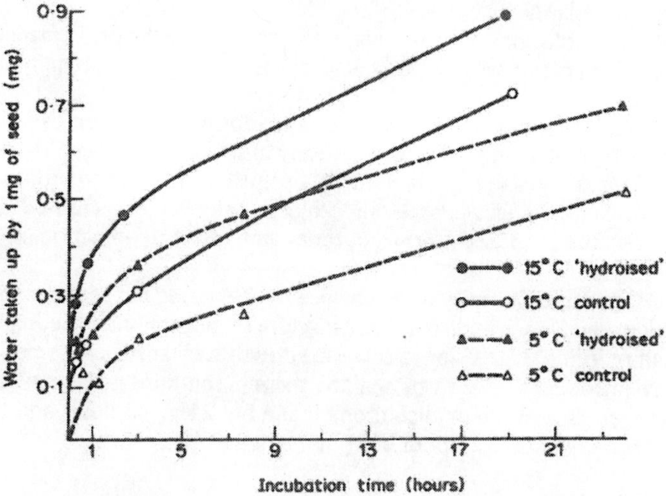

Figure 1.5 Imbibition of water by 'hydroised' and control carrot seed at 5°C and 15°C (after Austin *et al.*, 1969).

Limiting the progress of germination to a stage prior to radicle emergence can also be achieved by immersing the seed in solutions which osmotically limit the amount of water that can enter the seed. For example, soaking tomato seed in 1 to 2 per cent potassium phosphate plus 0.5 to 2 per cent potassium nitrate for 5 to 8 days followed by drying back enhanced the rate of field emergence of the seed. Polyethylene glycol (PEG) '6000' has been extensively used as an inert osmoticum for **priming** seeds and current research is largely concerned with defining the properties of seed lots that are the prerequisites of responsiveness to treatment, with determining the optimum temperature and osmotic conditions, and with investigating how to retain the maximum effect when the seed is dried.

Seed can be advanced to the point of emergence of the root and then sown using the technique known as **fluid-sowing**. The seed may be pre-germinated on wet paper towels, or in constantly aerated water. The latter technique can be used following priming and is used in large-scale commercial operations. Germinated seed may be held for 2 to 3 weeks at around 0°C if soil conditions make it impossible to sow. It is sown by mixing the seed in a gel of sufficient strength for the seed to remain suspended. This gel seed mixture is then extruded into the seed furrow.

On a small scale satisfactory results can be obtained using half-strength wallpaper paste (ensuring it does not contain a fungicide) as the carrier. The seed gel mixture can be extruded from a polyethylene bag which has been twist sealed and after filling has had a small corner removed to act as the nozzle. The aim should be to use about 10 ml of gel per metre of row. Too much gel can be detrimental as it excludes air.

Chemical stimulants

Pre-soaking in solutions of some chemicals can enhance the performance of seed. Soaking cotton seed in boric acid meets the high boron requirement of this crop.

The inhibition of growth due to seed dormancy can be overcome by environmental manipulation, as already discussed (see page 2), or in some instances by soaking the seed in a solution of a growth-promoting substance. Soaking lettuce seed in kinetin solution has ensured that high-temperature induced dormancy does not affect germination in the field.

The rate and uniformity of emergence of celery seed has been enhanced by soaking seeds in a growth regulator mixture of gibberellins and ethephon and then redrying them prior to sowing. Growth substances can sometimes be safely introduced to seed by soaking them in solutions made in organic solvents such as acetone or dichloromethane for 24 to 48 hours and then allowing the solvents to evaporate off immediately.

Seed ripening

The performance of seed can be affected by its environment while still attached to the parent plant. Low temperatures during seed development result in vernalisation, and the nutrition of the parent plant can affect the subsequent seed performance.

Although it is widely assumed that cultural and environmental factors during growth and ripening of vegetable seeds can influence their germination characteristics, data illustrating precise effects are sparse. It has been shown that carrot seed did not germinate when harvested 30 days after anthesis but 30 per cent germinated when harvested at 40 days and 93 per cent at 70 days after anthesis. The indications are that the embryo in carrot seed develops late and that seed should be left on the plant as long as possible to ensure the fullest possible growth of the embryos. When onion seed was ripened on the plant at 10 to 20°C and 20 to 30°C, the seed from the higher temperature gave an 83 per cent germination at 20°C as opposed to 76 per cent for seed from the lower seed-ripening temperature. The seedlings from the higher temperature seed had shoots 9.5 per cent larger and roots 31 per cent larger than those from low-temperature seed. These differences were not associated with differences in seed weight.

SEED STORAGE

Seed of cereals and pulses have always been important to agricultural man as sources of easily transported and stored food. Such seed, termed **orthodox seed**, can be dried to a low moisture content (5 per cent or less) without damage and then stored at normal temperatures for several years without losing its essential properties. Yet there are seeds which die rapidly in these conditions. These are termed **recalcitrant seeds** and their characteristic is that they cannot be dried below a relatively high moisture content without their doing irreversible damage. They can only be stored in moist conditions for relatively short periods. Large seeded woody species, such as beech, chestnut, elm, hickory, horse chestnut, poplar, walnut and willow, are those that typically have recalcitrant seed. Recent work has shown that seed of citrus, which was formerly classified as recalcitrant, may behave as orthodox if the testa is first removed.

Moisture content

The life span of most seed is increased if it is stored dry. The amount of moisture in a sample of seed is determined by drying a previously weighed sample (4 to 5 g) in a ventilated oven at 130°C for 1 or 2 hours and then reweighing after cooling in a desiccator for 30 to 45 minutes. The percentage moisture content is then calculated

$$\frac{\text{loss in weight of the sample}}{\text{original weight of the sample}} \times 100$$

This is sometimes referred to as the percentage moisture content on a wet weight basis and is the usual way of expressing results. This can be misleading in some respects. For example, in the right hand side of figure 1.6, seed lots with different moisture contents are represented diagrammatically, the weight of *dry* seed in each case being identical. Seed

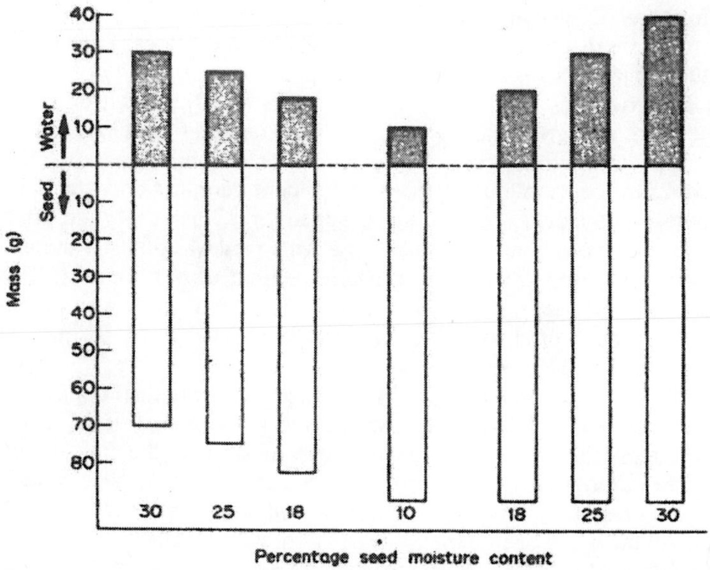

Figure 1.6 Diagrammatic representation of seed moisture contents; for discussion see text.

with a moisture content of 30 per cent contains four times as much water as seed with a 10 per cent content—*not* three times as much. Alternatively, in the left hand side of figure 1.6 the total amounts of *moist* seed are similar at each moisture content, and 100 minus the percentage moisture content gives a measure of the commercial worth of the seed; one hundred units of seed with a 10 per cent moisture content gives 90 units of dry seed, one with 18 per cent moisture gives 82 units and so on. It follows that seed with a 30 per cent moisture content should cost only seven-ninths as much as seed having a 10 per cent moisture content, provided that a high moisture content does not detract in some other way. Using more realistic figures, it can be calculated that seed with a 10 per cent moisture content should sell for just over 5 per cent less than seed with a 5 per cent moisture content. Considerable sums of money can be involved when bulk purchases of expensive seed are made.

Air-dried seed will usually contain 5 to 10 per cent moisture, the actual content depending on the species and the fluctuations in the moisture content of the air in which the seed is held. For example, after 105 days at 10°C and a relative humidity of 76 per cent, tomato seed had a moisture content of 16 per cent, lettuce 20 per cent, onion 30 per cent and peanut 14 per cent. The seasonal changes in relative humidity encountered in many climates could result in unsatisfactory moisture contents for much of the time.

Seed stores

The purpose of even the most primitive of seed stores is to protect dry seed from these fluctuating and high moisture conditions. Usually seed moisture contents of 4 to 6 per cent will preserve the viability of seed for more than twice as long as moisture contents of 10 per cent or greater. Furthermore at a given moisture content viability is retained longer at low rather than high temperatures. For example, onion seed held at a 12.5 per cent moisture content initially gave 98 per cent germination; after 1 year at laboratory temperature this had dropped to 82 per cent, and after 3 years to 1 per cent. When a sample of the same seed was similarly stored in the laboratory but with a lower moisture content of 6.3 per cent, it was still giving 71 per cent germination after 10 years but had lost all viability after 15 years. Stored at −4°C identical seed was giving 91 per cent germination after 15 years' storage.

How long seed will remain alive thus depends on the conditions of storage and perhaps above all else on the species of seed. This latter point can be emphasised by citing the example of seed of the Indian Lotus (*Nelumbo nucifera*) which has a particularly hard seed coat. Viable seed of this species has been found in an old lake bed where radio-carbon dating indicated that it had been buried for over a thousand years.

Stores used to conserve seed of material required as a genetic resource for plant breeding are usually held at −20°C with the seed at a low moisture content in vapour-proof packs. Orthodox seed is likely to remain highly viable for at least 30 years under such conditions. Some experimentation has been done on storing seed in liquid nitrogen at −196°C and it is claimed to be likely to give even longer storage at lower cost.

Modern materials and techniques have given us seed packets that are moisture proof. Packed in a very dry condition, seed retain their viability in these packets for several years even at ambient temperatures. However, once the packet is opened the seal is destroyed and deterioration will proceed unchecked.

The performance of stored seed

Maintenance of viability is not necessarily the same as maintenance of ability to establish plants or the maintenance of yielding capacity. The results of experiments on this topic again indicate that there is variation according to species, but the general picture is as presented graphically in figure 1.7. The time scale varies according to the conditions of storage and the species but it is fortunate that with many species similar performance can be obtained from well-stored old seed as from fresher seed. However, in seed lots where most of the viability has been lost the remaining seed may be genetically atypical of the original population and hence not give a similar performance. This is particularly important when seed is being conserved as a genetic resource.

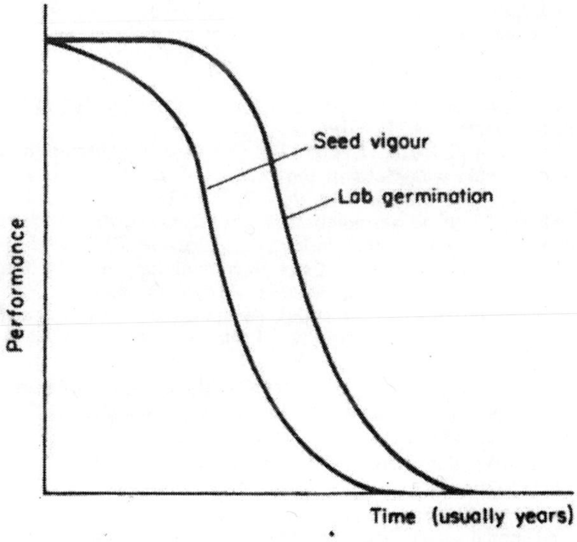

Figure 1.7 When seeds are stored their vigour deteriorates before their laboratory germination.

The process of ageing involves deterioration in many systems within the seed. Early in ageing the tissues become more leaky, indicating membrane damage. This is followed by loss of activity of some enzyme systems and loss of other cellular mechanisms. The final phase before death commonly involves genetic damage but some authors claim that there is some genetic damage at early stages of ageing.

There is some evidence that ageing in seed can be reversed by moistening them for a short period (one day) to allow repair processes to proceed before ageing damage becomes irreversible. This seems to parallel the enhancement of vigour associated with priming. It seems likely that different species will lose viability for different reasons. Oily seed such as those of cotton or flax lose viability because the oil breaks down into free fatty acids; chemicals which inhibit this process can help to retain the seed viability. The accumulation of toxic metabolites or the loss of nuclear structure or life processes are slowed to a minimum in good storage conditions. Thus the seed can retain for many years its essential capability of germinating to give the seedling.

References

ANON. (1976). International rules for seed testing 1976. *Seed Sci. Technol.*, 4, 1-180.

AUSTIN, R. B., LONGDEN, P. C., and HUTCHINSON, J. (1969). Some effects of 'hardening' carrot seed. *Ann. Bot.*, 33, 883-95.

COURTNEY, A. D. (1968). Seed dormancy and field emergence in *Polygonum aviculare*. *J. appl. Ecol.*, 5, 675-84.

CRAIN, D. E. C., and HAISMAN, D. R. (1963). A note on the slow re-hydration of some dried peas. *Hort. Res.*, 2, 121-5.

GREGORY, F. G., and PURVIS, O. N. (1938). Studies in vernalisation of cereals; III The use of anaerobic conditions in the analysis of the vernalising effect of low temperature during germination. *Ann. Bot.*, 2, 753-61.

HARRINGTON, J. F. (1960). Germination of seeds from carrots, lettuce and pepper plants grown under severe nutrient deficiencies. *Hilgardia*, 30, 219-35.

HEWSTON, L. J. (1964). Effects of seed size on germination, emergence and yield of some vegetable crops. *M. Sc. Thesis, University of Birmingham*.

JENSEN, H. A., and JØRGENSEN, J. (1969). The influence of the degree of maturity and drying on the germinating capacity of *Festuca pratensis*. *Huds. Acta Agric. scand.*, 19, 258-64.

MATTHEWS, S., and BRADNOCK, W. T. (1967). The detection of seed samples of wrinkled-seeded peas (*Pisum sativum*) of potentially low planting value. *Proc. int. Seed Test. Ass.*, 32, 553-63.

PERRY, D. A. (1981). Report of the vigour test committee 1977-1980. *Seed Sci. Technol.*, 9, 115-26.

SHAPLEY, S. C. C., and VARNER, J. E. (1970). Respiration and protein synthesis in dormant and non-dormant seed of *Avena fatua*. *Pl. Physiol., Lancaster*, 46, 108-12.

Further Reading

GRAY, D. (1981). Fluid drilling of vegetable seeds. *Hort. Reviews*, 3, 1-27.

HEBBLETHWAITE, P. D. (Editor) (1980). *Seed Production*. Butterworths, London.

HEYDECKER, W. (Editor) (1973). *Seed Ecology*. Butterworths, London.

HEYDECKER, W., and COOLBEAR, P. (1977). Seed treatments for improved performance — survey and attempted prognosis. *Seed Sci. Technol.*, 5, 353-425.

ROBERTS, E. H. (1981). Physiology of ageing and its applications to drying and storage. *Seed Sci. Technol.*, 9, 359-72.

CHAPTER 2

THE SEEDLING

The growth of any living thing ultimately depends on photosynthesis – the process by which green plants use the sun's energy to build complex organic chemicals from simple inorganic substances. Even the carnivorous animal indirectly depends on photosynthesis. The carnivore's prey will either have been a herbivore or itself a carnivore. For all these food chains, the primary source of new energy is the sun and this energy is harnessed by plants through photosynthesis.

Fungi are plants lacking the green pigment chlorophyll which is vital for photosynthesis. They have to obtain complex organic compounds either from other living material, in which case they are *parasites*, or from the decaying debris of once living things, in which case they are *saprophytes*.

The germinating seed at first depends on its reserves of food, derived from the photosynthesis of the parent plant. Before this store of food is exhausted, the seedling must begin its own photosynthesis if its growth is to continue. What is the nature of the chemistry of photosynthesis?

PHOTOSYNTHESIS

The overall reaction involved in photosynthesis is often represented by the equation

$$CO_2 + H_2O \xrightarrow{\text{light energy}} (CH_2O) + O_2$$

This equation wrongly suggests that the oxygen released by the plant is derived from the carbon dioxide: it is now known that the oxygen is derived from water. Thus the equation becomes

$$CO_2 + 2H_2O \xrightarrow{\text{light energy}} (CH_2O) + O_2 + H_2O$$

and is known as the *Van Niel equation*. Both equations reveal however that the energy of light is used within the plant to convert carbon dioxide and water to more complex organic molecules such as glucose, $(CH_2O)_6$.

Green plants reflect green light and absorb the other colours that make up 'white' light. Reflection of green light is not complete and some is absorbed, so we cannot assume that the green wavelengths play no part in photosynthesis. Balegh and Biddulph (1970) were the first to use sophisticated equipment to determine the relative importance of the different wavelengths of light for photosynthesis in a higher plant. They investigated the photosynthetic efficiency of light having a wavelength of from 400 to 700 nanometres, using increments of 12.5 nm and leaves of *Phaseolus vulgaris* cv. Red Kidney. Normally we think of light as having a continuous wave form but light energy is in discrete units termed **photons**, each of which 'carries' one quantum of energy. We can measure the rate of photosynthesis by measuring the rate at which carbon dioxide is absorbed by leaves and calculate the effectiveness of a particular wavelength in terms of the number of molecules of carbon dioxide absorbed per 1000 photons of light incident on the leaf. Their results are summarised in figure 2.1 and reflect the activity of a number of pigments, which can be regarded as **catalysts** that are not changed by the photochemical reaction. The main active pigments are

Chlorophyll a, which is blue-green with principal absorbing peaks at about 430 and 660 nm.

Chlorophyll b, yellow-green with peaks at 465 and 660 nm.

Phycocyanins, which form a group of blue pigments with absorption in the 560 to 660 nm region of the spectrum.

Carotenoids, which form another group of red, orange, yellow or brown pigments. They are divided into two chemical groups, the carotenes and the xanthophylls, and both absorb in the 400 to 500 nm part of the spectrum.

The interplay of such pigments results in many different action spectra. Figure 2.1 is probably a typical example of higher plants.

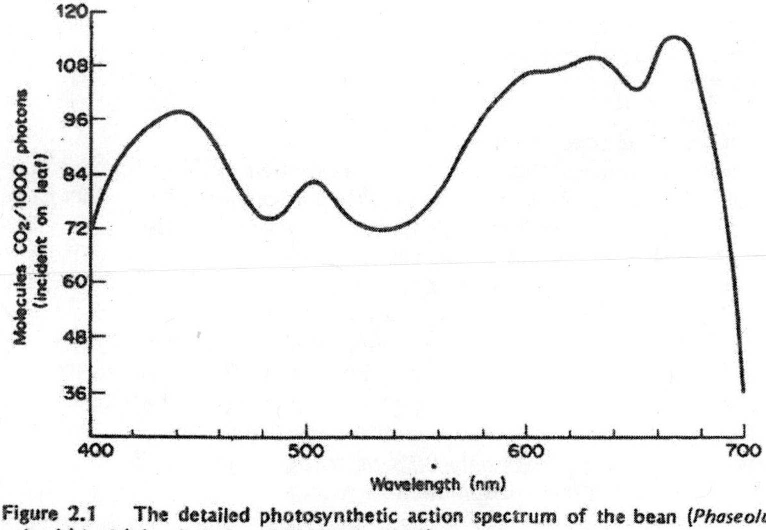

Figure 2.1 The detailed photosynthetic action spectrum of the bean (*Phaseolus vulgaris*) leaf (after Balegh and Biddulph, 1970).

We can conclude, so far, that the rate of photosynthesis depends on the light energy available, the effectiveness of the pigments in absorbing and transferring this light energy and the availability of carbon dioxide.

Theoretically the photochemical reactions in photosynthesis require about four quanta per molecule of carbon dioxide absorbed. Measurements using the leaves of higher plants in ideal conditions indicate that nine to thirteen quanta are needed. This is the **quantum number** and measures the efficiency of photosynthesis in the leaf. The range of nine to thirteen quanta indicates efficiencies ranging from about 30 to 45 per cent. This efficiency can also be expressed as molecules of CO_2 fixed per quantum (the **quantum yield**), which for the range cited would be 0.11 to 0.08. Measurements in a crop of maize have shown that quantum yields as high as 0.064 can occur for limited periods.

It must be remembered, however, that light effective for photosynthesis (in the 400 to 700 nm wavelength range) comprises only some 47 per cent of the total radiation reaching the earth from the sun. This total radiation has a maximum intensity, known as the **solar constant**, of about 6.7 J cm^{-2} min^{-1}. The efficiency of photosynthesis in relation to total radiation drops to about half the figures so far quoted and we are left with an efficiency of about 15 to 22 per cent.

Although this value may seem low, it must be compared with the efficiencies of about 1 per cent for crops and natural vegetation measured over the year. Very often only a small part of the total production is usable, reducing the overall efficiency of photosynthesis to as little as 0.2 per cent. The scope for improvement seems so great that there is a temptation to believe that higher efficiencies can easily be achieved. In practice many factors combine to limit photosynthetic efficiency and some of these of particular relevance to the horticulturist will now be discussed.

Carbon dioxide concentration

Normal air contains about 300 parts per million of CO_2. Some of this enters the leaf via the stomata and makes contact with the photoreactive pigments in the chloroplasts. Initial entry is achieved by gaseous diffusion but to reach the chloroplasts the carbon dioxide dissolves and diffuses in the cellular water. During diffusion there is a 'flow' along a chemical gradient, from a high to a low concentration. Because the plant uses carbon dioxide the concentration around it is reduced, thereby increasing the 'driving force' for the diffusion to the plant of further carbon dioxide. In still air this effect reduces the concentration of carbon dioxide in a 1.3 cm layer above a continuous crop surface. Wind reduces the height of this layer, which becomes only 0.38 cm if the windspeed is 8 km h^{-1}. As wind flows over a leaf, streamline flow occurs in a layer close to the leaf surface, called the boundary layer. The thickness of this boundary layer decreases with increase in windspeed but increases with leaf area. Therefore, this external resistance to diffusion is not eliminated by wind, albeit markedly reduced. Compared with the internal resistances to carbon dioxide flow, this is negligible; although the distances that carbon dioxide has to diffuse within the leaf are small, the rate of diffusion in water is about 10 000 times slower than in air. Heath (1969) calculated that for a typical mesophyte leaf, like that of *Pelargonium zonale*, the resistances to carbon dioxide utilisation are partitioned as in figure 2.2.

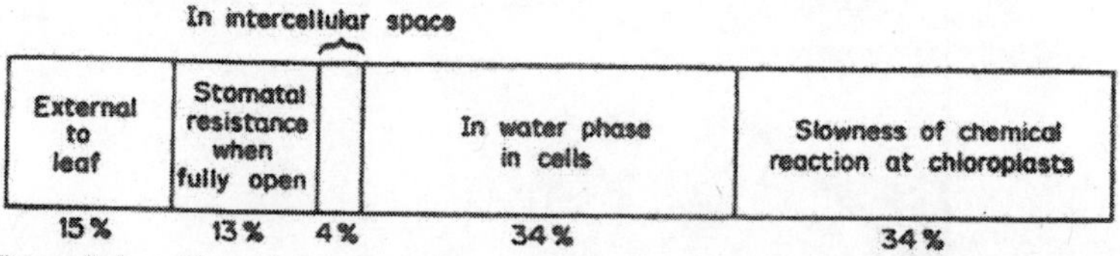

Figure 2.2 The relative resistances to carbon dioxide utilisation by a green leaf, fully illuminated, with its stomata fully open in a light breeze (based on Heath, 1969).

This partitioning is for a fully illuminated leaf with its stomata fully open in a light breeze. If the stomata close to one-third the aperture, the total resistance to flow is approximately doubled and the resistance of the stomata to carbon dioxide entry increases from 13 per cent of the total to about 55 per cent.

Further closure of the stomata increases the resistance dramatically and it is evident that for photosynthesis to proceed at optimal rate full stomatal opening is needed.

CO_2 enrichment

The rate at which carbon dioxide diffuses to the sites of photochemical reaction has so far been discussed in terms of a series of external and internal resistances to flow. The overall impetus to diffusive flow is however a function of the *difference* in concentration between the air and the sites of photochemical reaction. It follows that we can increase the rate of diffusion by increasing the concentration of carbon dioxide in the air. A new commercial practice artificially increases the carbon dioxide concentration in glasshouses. This will now be considered in some detail because of its horticultural significance.

Consider the general principles involved. We are interested in the manner in which the rate of photosynthesis is affected by the concentration of carbon dioxide. To make the case general, let us say we are interested in how the rate P of a process is affected by the concentration C of a substance. We might expect that P will consistently increase with increase in C, a simple example being represented graphically in figure 2.3 by the line AB. If the rate P were however also affected by something else (by analogy, a man may be unable to lay bricks any faster than a certain

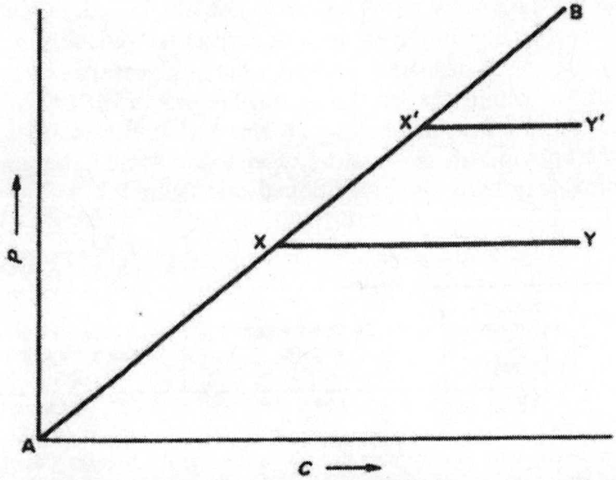

Figure 2.3 The rate *P* increases with increase in the supply of *C* in a manner represented by the line AB. However, if *P* also depends upon the supply of some other factor(s) in addition to *C*, then *P* will not continue to increase with *C*. Curves approximating to AXY or AX'Y' will result.

rate, not because he is short of bricks, but because he is short of mortar), *P* can only increase with *C* to a certain point; thereafter any increase in *C* has no effect on *P*. This is represented by AXY in figure 2.3. If in the analogy the supply of mortar is increased, then once again the rate at which bricks can be laid will depend on their supply, until the mortar supply again limits the overall rate (line AX'Y' in figure 2.3).

When several factors are involved in a process the rate of the process is limited by the factor in least abundant supply. There are several factors involved in photosynthesis but it will suffice to consider two of the most important ones – the concentration of carbon dioxide and the amount of light falling on the leaf. If light intensities are low, then increasing the carbon dioxide concentration will not increase the rate of photosynthesis.

Figure 2.4 shows that there is no photosynthesis in the dark. At the lowest light intensity I_1 the 300 ppm of CO_2 in normal air limited photosynthesis but an intensity I_3 permitted the increase in photosynthesis rate as the CO_2 concentration increased up to 600 ppm, and I_4 up to 900 ppm of CO_2.

The family of curves in figure 2.4 can tell us a great deal; for example, there is no virtue in enriching glasshouses with CO_2 at night. Second, the maximum benefits from CO_2 enrichment are not likely to be obtained in poor light – CO_2 enrichment should be linked to light intensity. In practice, fluctuations in CO_2 concentration may not be practical because of the high cost of control. Provided the glasshouse does not leak too much, it may be better to maintain a high concentration to take advantage of even short bursts of bright sunlight. Lettuce make particularly effective

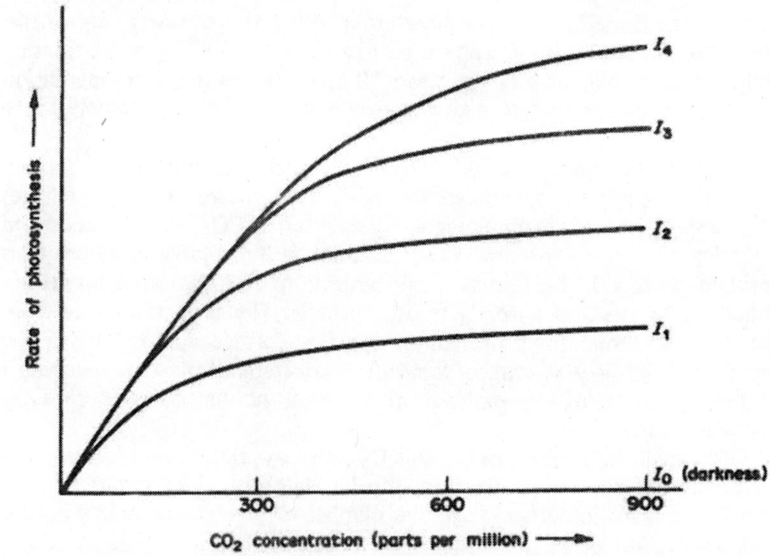

Figure 2.4 A diagrammatic representation of the relationship between the rates of photosynthesis and carbon dioxide concentration at a series of increasing light intensities (I_0 –I_4).

use of very low light intensities, so with this crop CO_2 enrichment can be worthwhile in the dull days of a temperate winter.

Experiments at the Glasshouse Crops Research Institute, Sussex, England, have indicated that maintaining a concentration of 1000 vpm ($2.0\,g\,m^{-3}$) throughout the life of a tomato crop (20 weeks) is economic. The yields of ripe fruit after 4 weeks' picking from a plant in air were 0.32 kg; in 600 vpm CO_2, 0.89 kg; in 1000 vpm CO_2, 1.07 kg; and in 1400 vpm CO_2, 1.13 kg. The corresponding total yields after 20 weeks were 4.36, 5.50, 5.97 and 6.07 kg. Clearly the greatest effects were on early yield, which commands a high price in the UK. Furthermore, the diminishing benefit derived from increasing concentrations illustrates the operation of a limiting factor. Such factors rarely produce the abrupt limitation of increase illustrated in figure 2.3. It is more usual to find a gradual diminution of the increase until a relatively stable rate is reached, as is indicated in figure 2.4.

Photorespiration

In a sealed system (such as a bottle garden) or in some inadequately ventilated growth cabinet, plants growing in the light will deplete the air of carbon dioxide. The concentration of carbon dioxide will eventually stabilise at the CO_2 **compensation point.** At this point the amount of carbon dioxide used in photosynthesis is equal to that produced by respiration. Experiments have shown that with plants of maize, sugar cane, sorghum and several other tropical crops and weeds, the CO_2 compensation point is very low, usually less than 10 ppm. However the compensation point for most temperate higher plants is usually greater, namely 50 to 100 ppm.

The higher compensation point in temperate plants results from a type of respiration stimulated by light. This *photorespiration* releases CO_2 inside the leaf, maintaining a higher internal CO_2 concentration and reducing the influx of new CO_2. Consequently a stable equilibrium is reached with a higher external concentration of CO_2. Such plants are called C_3 plants and have C_3 photosynthesis. The term is derived from the three carbon atoms present in the first-formed compounds of the metabolic pathway of carbon fixation. Photorespiration is a direct result of the operation of this pathway at or below normal atmospheric CO_2 concentration.

Tropical plants also possess this C_3 pathway, but many of them have an additional system of fixation, the C_4 pathway which precedes this. The name is again derived from the number of carbon atoms in the first metabolic product of this pathway. The C_3 pathway in C_4 plants occurs at some distance from the sub-stomatal cavity and the products of C_4 fixation are transported to this site. Here CO_2 is released for re-fixation at concentrations high enough to inhibit photorespiration. In addition to this C_4 plants have a much greater affinity for CO_2 and can thus maintain low concentrations of CO_2 in the sub-stomatal cavity. This results in better diffusion gradients being maintained when the stomata are partially closed to prevent transpiration and can be seen as an adaptation to bright light and arid conditions.

Net photosynthesis – that not burned up by respiration – is usually measured in terms of net CO_2 uptake per unit weight or area of leaf. From what has already been said, it will be apparent that in high intensity light the net influx of CO_2 will be greater, other conditions being equal, the lower the CO_2 concentration inside the leaf.

This simplified model suggests that plants without photorespiration should be more efficient by virtue of having a greater net influx of CO_2. Short-term measurements indicate that this is so, but it does not necessarily follow that these plants make more efficient crops. Growth pattern and form can play a considerable part in reducing the impact that photorespiration may have.

Light

Light intensity is a major factor governing the rate of photosynthesis. The net photosynthesis of a single leaf that is fully supplied with water will increase with increasing light intensity until some limiting factor causes a more or less steady rate to become established (see figure 2.3). This second factor is normally the CO_2 content of the air, but we have seen that by increasing the CO_2 content a higher level of photosynthesis may be attained. The maximum *efficiency* of light energy conversion is represented however by the steepest part of the curve in figure 2.3, which is in the part where light intensity is low. At these low intensities a single leaf may use up to about 13 per cent of the incident energy in the 400 to 700 nm waveband. In high light intensities this efficiency can drop drastically as photosynthesis becomes limited by some factor other than light. A plant will use light more efficiently if its leaves are evenly illuminated at low intensities, than if some of them were over-saturated by high intensities and others in deep shade.

The arrangement of leaves can play a part in determining the photosynthetic efficiency of a crop canopy, as shown in figure 2.5. The highest rates of assimilation by crops are not encountered in seedlings, where there is relatively little leaf per unit area of ground. The leaf area/ground area ratio is widely used in studies to determine the pattern of growth of crops and is termed

$$\underset{\text{LAI}}{\text{Leaf Area Index}} = \frac{\text{Area of leaf}}{\text{Area of ground}}$$

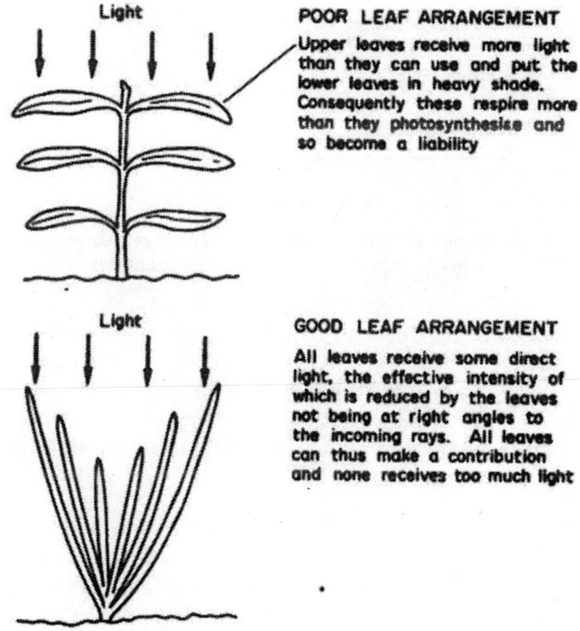

Figure 2.5 Arrangement of leaves.

It is convention to consider the leaf area as the plan area (one surface) and not the surface area (both surfaces).

At low LAIs much of the light energy hits bare soil or is wasted because the leaf is receiving more light than it can use. The **optimum LAI for a crop** gives the highest rate of assimilation per unit area of ground (see chapter 3) and in practice curves such as figure 2.6 are obtained. This indicates that about five layers of leaf are required before the canopy of the hypothetical crop reaches its maximum efficiency; experimental values between three and eight are commonly obtained. At values higher than the optimum the assimilation per unit area of ground is lower because some leaves are in deep shade and respire more than they photosynthesise.

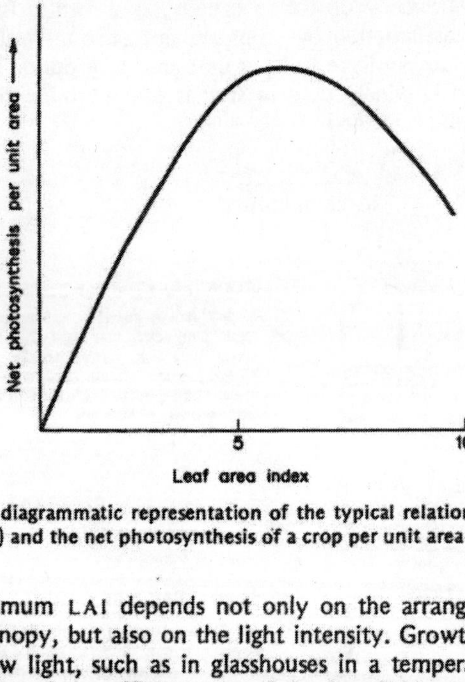

Figure 2.6 A diagrammatic representation of the typical relationship between leaf area index (LAI) and the net photosynthesis of a crop per unit area of ground.

The optimum LAI depends not only on the arrangement of leaves within the canopy, but also on the light intensity. Growth will be slow in periods of low light, such as in glasshouses in a temperate winter. This occurs in spite of the efficient use of the low light and the artificially maintained high temperatures, but artificial light can be used to increase photosynthesis.

Supplementary lighting has the potential of increasing the early yield of glasshouse crops. This practice is however very costly and not all sources of visible light are equally suitable. Tubular or bulb discharge lamps are favoured because they give more light per unit of electricity than other kinds. Discharged electrons impart energy to gas molecules inside the tube or bulb by collision. This energy is rapidly emitted as light of wavelengths depending on the gas and its pressure in the tube. The most commonly used gases are mercury vapour, sodium vapour and the inert gases (such as xenon). Some of the wavelengths that these gases produce are outside the visible range and therefore are no use for lighting or photosynthesis. However, these useless wavelengths can be converted to visible light by coating the inside of the tube or bulb with fluorescent material. This emits visible light when it is energised by light of invisible wavelengths.

The ordinary tungsten-filament light bulb does not produce visible light very efficiently as much of the electricity consumed is converted to heat. If high intensities of visible light are achieved using such tungsten bulbs, then the associated intense heat is normally detrimental to plants. As we shall see later, this form of lighting does have other important uses in horticulture but it is not a good source of photosynthesis. A good source would be one with a spectral composition similar to that suggested by figure 2.1, namely one with peaks of emission at 400 to 500 and 600 to 700 nm.

Different criteria can be used to compare the value of different light sources. Simple measurements of the intensity of light in the 400 to 700 nm band are used to make comparisons of the efficiency of different lamps. For example, mercury fluorescent reflector lamps (MBFR/U) emit 14.9 per cent of the energy supplied as useful light in the 400 to 700 nm band. The more modern high-pressure sodium lamps (Type SON) have a corresponding efficiency of 27 per cent.

Point sources of light suspended above a glasshouse bench give intense illumination immediately beneath and progressively less intense illumination to the sides. By proper spacing at a set height (usually recommended by the manufacturers), the lower intensity areas of two adjacent lamps can be made to overlap and produce a more uniform illumination over the whole area. This is important because uniform batches of plants can only be raised by having uniform lighting.

As mentioned previously, supplementary lighting can be used to obtain earlier crops by encouraging seedling growth in the dull days of a temperate winter. Several workers have shown the advantages of doing this for glasshouse tomato crops in the UK and typical results are shown in figure 2.7.

Almost any plant will respond favourably to supplementary lighting if the natural light intensities are low, but an economic return usually depends on treating many plants per lamp for about 2 to 4 weeks to obtain earlier or improved yields sometime later.

It is usual to illuminate the plants for 12 hours each day. One set of lights can be moved (or the plants!) to double their use each 24 hours. This has been shown to work well on tomatoes and some other crops, but there are some species which give "drawn" plants under some light sources if they do not get a period of unsupplemented daylight before their night. Tomatoes become chlorotic and even die if they are in continuous light,

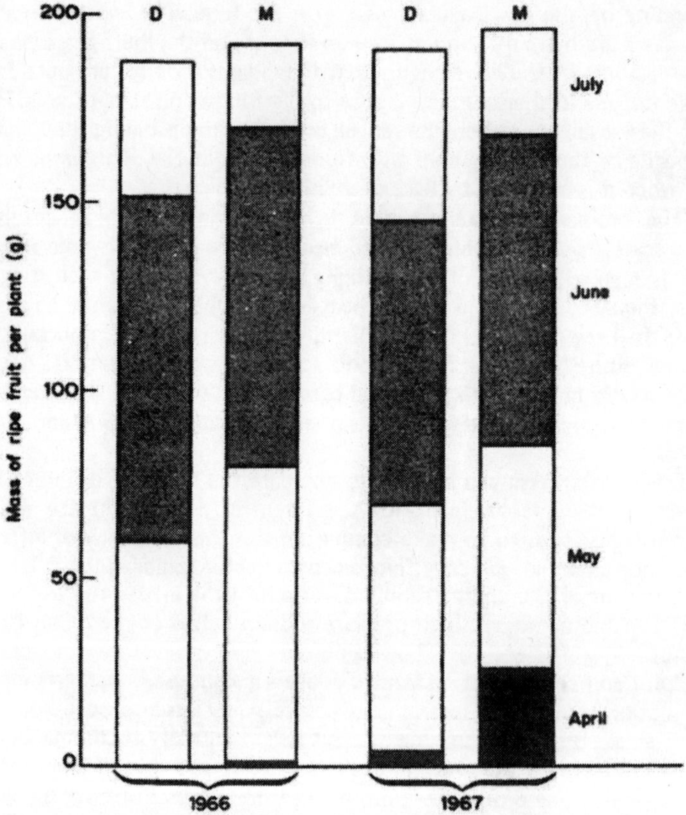

Figure 2.7 The effect of supplementary lighting for tomato seedlings (cv. Moneymaker) on subsequent cropping. D, seedlings grown in daylight only; M, seedlings receiving 3 weeks for supplementary mercury light for 12 hours each day (data from A. E. Canham, Dept. of Horticulture, University of Reading).

and the development of many species (particularly the flowering) can be affected by the day-length (see chapter 5).

In spite of these provisos, supplementary illumination can improve a seedling's start in life and even manifest itself when the crop is harvested. We shall return to discuss the circumstances in which this is so and those in which it is not, when we consider seedling growth towards the end of this chapter and growth analysis in chapter 3. For the present we return to the processes of growth and in particular to the water relations of the seedling.

WATER RELATIONS

Between about 80 and 95 per cent of the fresh weight of herbaceous plants is water. In chapter 1 we saw how hydration of the colloids and activation of enzymes depend on water uptake by the seed. The continuing growth of the seedling depends equally on water. In a typical higher plant water is obtained from the soil; about 98 per cent of this is lost again by the diffusion of water vapour through the leaf stomata and to a lesser extent through the cuticle.

Transpiration

Water vapour diffuses out of open stomata by a process called transpiration. Because of the differences in the nature of the resistances to this diffusion as compared with carbon dioxide diffusion, progressive stomatal closure reduces water loss more than photosynthesis. Thus, partial closure of stomata is an effective way of reducing water loss with a minimal impact on photosynthesis. Nevertheless stomatal closure due to excessive water stress does limit photosynthesis, especially in bright sunlight capable of giving maximum photosynthetic rates.

It might be horticulturally advantageous to apply water to plants and so avoid the temporary wilting and stomatal closure, which often occurs in the field and glasshouses around noon. However, experiments designed to test if increased yields could be obtained by spraying small quantities of water on to crops to avoid wilting have not consistently given increased yields. This may be due to limitations of the experiments and should not at this stage be taken as indicating that the likely effects of such temporary wilting can be ignored.

Wilting and stomatal closure can occur when the roots are in wet soil, because the roots cannot supply water fast enough and an internal water stress develops, often termed *physiological drought*. It is commonly seen in understocked glasshouses where the absence of other transpiring plants and the high temperatures reduce relative humidity to very low levels which accelerate the diffusive loss of water through the stomata.

Stomatal aperture

As previously mentioned, the stomata of the leaf must be open during periods of light to allow carbon dioxide for photosynthesis to enter. Carbon dioxide has to dissolve in water to reach the sites of the photochemical reaction; wet surfaces are therefore evaporating water into the spaces in the mesophyll and then through the stomata to the outside air. The rate of this inevitable water loss from the sub-stomatal cavities depends on factors controlled by the plant and on the environment.

Chief among these is the *stomatal aperture*, which is under multi-factor control. One major controlling factor is the concentration of carbon dioxide in the sub-stomatal cavity. In darkness the concentration is increased by the CO_2 from respiration. In light, photosynthesis uses up this CO_2 and the stomata open. Stomata can be made to open in the dark by putting plants in CO_2-free air, so that the internal CO_2-level is reduced by diffusion rather than photosynthesis. Conversely, closure can occur if CO_2 levels in the sub-stomatal cavity are increased. This has been shown to happen not only in darkness but also when the temperature is high. It also appears that water stress, although in itself not inducing sufficiently high CO_2 concentrations to close stomata, may cause stomata to close by inducing a greater sensitivity to CO_2 concentration. How internal CO_2 concentration affects stomatal aperture is not known, but it may be that the reduction of starch in the guard cells which accompanies opening in both light and dark will prove a useful clue.

Water stress has been shown to increase the level of the plant growth regulator abscisic acid (ABA) in leaves and when ABA is applied to leaves stomata close rapidly and remain closed. However, it is questionable whether stomatal closure is directly (primarily?) regulated by ABA as stress closure may occur before the ABA concentration increases measurably. Furthermore, when leaves have regained full turgidity after a period of water stress, their stomata open while ABA levels are still high. Thus it seems improbable that ABA plays a primary role in the regulation of stomatal aperture.

That our knowledge of the control mechanisms is inadequate can be further illustrated by the fact that high temperatures can induce stomatal opening if they are not associated with a rise in internal CO_2 concentration. Whatever the stimuli there is some evidence that they are transmitted to other parts of the same leaf – or even to other parts of the plant – probably via the phloem.

This multi-factor control system means that stomata are not simply open during the day and closed at night. High temperatures around noon may cause closure quite independent of excessive water loss. Wind may remove some barriers to the diffusive entry of CO_2 and cause higher CO_2 concentration in the sub-stomatal cavity and, hence, stomatal closure. In neither of these two instances has regulation of transpiration been seen as the cause of stomatal closure.

Stomata usually close if an internal water shortage develops, but this does not fully prevent water loss. Provided there is no change in the external condition, the rate of water loss from a plant with its stomata closed depends largely on the thickness of its cuticle. In plants with a thick cuticle, transpiration with stomata closed may be as little as 2 per cent of that when the stomata are fully open. With much thinner cuticles this figure could be as high as 45 per cent. This considerable range of cuticular transpiration is associated with the adaptation of plants to their environment. In general, plants with a thick cuticle are better adapted to survive in dry environments than those with thinner cuticles. Dry conditions are found in deserts or during physiological drought, such as is encountered by evergreens in a temperate winter. Here low temperature limits water uptake and if cuticular transpiration were not low, there could be a net water loss.

Other morphological characters, particularly those of leaves, limit transpiration and characterise *xerophytes* (plants that can live in dry places); they include sunken stomata, a reduction of the leaf surface (seen at its extreme in the Cacti) and the presence of surface hairs. These characters probably play a part in reducing water loss during shortage, but when plentifully supplied with water, xerophytes transpire at rates as high, and commonly higher than, those of some *mesophytes*, which are plants adapted to average water supply.

Water uptake

One common xeromorphic character is an extensive and deep root system, enabling the tapping of a large volume of soil. Water is held in the soil by capillary and surface tension forces within the small spaces between soil particles and is adsorbed as a surface film around them. Plants take up water against these forces and raise some water to the upper leaves – a lift of some 100 metres in the tallest trees. Most plant physiologists agree that there is no fully satisfactory explanation of how this is achieved. Modern application of physical concepts will however probably give a greater understanding of this process.

Briefly, the passage of water from the external surface of the root to the xylem of the root can be explained by several satisfactory and not mutually exclusive hypotheses. The movement of water up the xylem is less satisfactorily explained by current theories, most of which consider transpiration as the source of energy for pulling the water to the highest point. This pull is transmitted by continuous columns of water within the xylem vessels. Water has sufficient cohesive strength to exist in such long columns without breaking up into shorter lengths; superficially, therefore, the theory is satisfactory. A continuous thread of water under tension in narrow tubes breaks up readily, however, if there is the slightest shaking or vibration. How do trees, shaken and vibrated by wind, maintain cohesion in the columns of water in their xylem vessels? The answer appears to be that the columns *do* break and air enters, but somehow repairs are made and water transport continues. Some believe the repairs to involve a redissolving of the air under the influence of **root pressure**. It has also been suggested that when the column of water within the xylem vessels breaks, continuous micropores within the walls of the xylem vessels take over so that transpiration pull can still satisfactorily account for the upward water movement.

There is good evidence that the roots are not passive in water movement but pump water up the xylem under pressure. The simplest evidence supporting the pumping action of roots is the existence of *root pressure*. This can be demonstrated on a grand scale by felling trees and measuring the pressure necessary to prevent sap from exuding out of the cut stump.

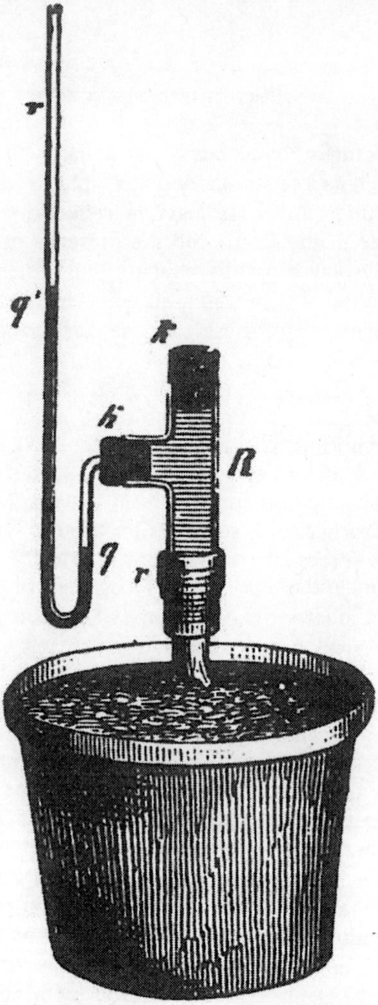

FIG. 438.—Apparatus for observing the force with which water escapes under root-pressure from the transverse section of a stem *r*. The glass tube *R* is first of all firmly fastened to the stem, and the tube *r* then fixed into it by the cork *k*. *R* is completely filled with water, the upper cork *k* then fixed in it, and mercury poured into the tube *r* so as to stand from the first higher at *q'* than at *q*, the level *q'* rising above *q* according to the intensity of the root-pressure. The apparatus is much more convenient to handle than that hitherto in use.

Figure 2.8 (Reproduced, complete with caption, from Sachs' *Textbook of Botany; Morphological and Physiological*, 1875 edition).

Smaller experiments with saplings or the obliging *Pelargonium* can be found reported in early textbooks on plant physiology (figure 2.8). Root pressures of 1 to 2 atmospheres have commonly been recorded and it has been suggested that they may abnormally reach as high as 7 to 8 atmospheres. To record root pressures there must be no shortage of oxygen, the temperature must not be low and the roots must come from a healthy well-nourished plant. What are the forces involved here?

Because we are interested in flow, our concern is with differences in free energy available to do work, and with the tendency towards equilibrium, in which all components have equal free energy states. If we dissolve a salt in water, the molecules of the water are held apart by the presence of the salt, so decreasing the free energy of the system.

The forces for the uptake of water into the vacuole of a normal plant cell represent its *water potential*. This results from the free chemical energy of the water in the vacuole, which exceeds that of pure water as a result of the excess pressure or constraint of the cell wall. The free chemical energy of the vacuole water is reduced by the presence of dissolved salts. This is usually written

ψ	=	P	−	π
(Psi — water potential)		(Pressure in excess of one atmosphere)		(Pi — the reduction in chemical energy due to solutes)

Water potential is expressed in terms of energy per unit volume (for example, erg cm^{-3}, joule cm^{-3}) or in terms of force per unit area (such as atmospheres or bars). Traditionally biologists have preferred to work in atmospheres and the osmotic pressure of the liquid in the vacuole is the term π in the above equation. An additional term can be introduced to the equation to take account of the reduction in the chemical potential of water that occurs at solid- and gas-liquid interfaces. Thus

$$\psi = P - \pi - \tau$$

(tau — the matrix potential)

The two terms π and τ must exceed P to give the water potential a negative value if water is to enter the system. The *water potential* is the difference between the chemical potentials of the system and pure water (at atmospheric pressure and the same temperature as the system). The negative gradient of the chemical potential drives water from outside the cell into the vacuole. Two systems, the water outside the cell and the liquid in the vacuole, are separated by a semi-permeable membrane, which lets water pass only; the rate of water movement is proportional to the driving force. The same holds for the flow of water between cells separated by semi-permeable membranes. The rate of flow is proportional to the difference between their water potentials, and movement will be towards the cell with the more negative water potential.

The soil water that is in contact with the absorbing surface of the roots moves across the root cortex, through the passage cells of the endodermis and through the pericycle parenchyma to the xylem vessels. It is not known how much of the total water movement is through rather than between these cells. It was usually held that the endodermis was the major constraint to flow, because the Casparian strip seemed to be designed to restrict water entry into the stele to the narrow pathway of the specialised passaged cells. These cells seemed to represent the chief bottleneck to the entry of water, but recent experiments have indicated that water entry into the xylem vessels themselves is the main resistance to flow.

The forces producing this flow and the factors governing its rate are all thought to be physical and should not be affected by metabolic inhibitors or temperature changes. But affected they are, and this lends support to the idea of a biological pump. However when the inter-relationships between salt uptake and water uptake are considered another explanation emerges.

SALT UPTAKE

Inorganic salts, such as the fertiliser potassium nitrate, are dissolved in the soil water and dissociated into ions. The forces enabling these ions to enter the plant are exactly the same as those that cause the entry of water; they are gradients of chemical potential, termed the **electrochemical potential.**

Plants concentrate the salts from the soil water and accumulate them in proportions different from those found in the soil. Thus ions are accumulated and selected as they pass through the membranes of the cells in contact with the soil water. Put very simply, the membranes are selectively permeable, letting some ions through more easily than others. Electron micrographs indicate that this selectivity is achieved by the sieve mesh of the membrane. The accumulation is *against* the natural energy flux, so the plant has to use some of its photosynthetic energy to accumulate salts and move them through the xylem transpiration stream to sites of metabolism.

The presence of salts within the cell sap lowers the chemical potential and increases the energy gradient – the water potential (see page 39). Because the accumulation of salts depends on respiration and because these salts affect the rate of water uptake, it can be argued that water uptake also requires energy from respiration. The actual uptake of water is thought however to depend entirely on physical forces; but the maintenance of a continuing uptake depends on salt uptake and hence respiration.

Essential elements

The elements acquired by the plant from the soil in relatively large amounts – the *macro-nutrients* – are nitrogen, phosphorus, potassium, sulphur, calcium and magnesium. Other elements from the soil are essential for plant growth but only very small quantities are needed. These are termed **micro-nutrients** or **trace elements**: iron, manganese, boron, zinc, molybdenum, chlorine and copper. The generally accepted criteria used in determining whether an element is essential or not were proposed by Arnon and Stout in 1939.

(1) The plant cannot grow or reproduce normally if the element is not supplied.
(2) The action must be specific (irreplaceable by some other element).
(3) The element must be directly required by the plant; it must not act, for example, by stimulating the uptake of some other element or combating the toxic effect of some substance.

Insufficient supplies of both macro- and micro-nutrients can produce characteristic symptoms in some plants. The 'whiptail' in cauliflowers (figure 2.9) is caused by a deficiency of molybdenum. Iron deficiency causes a yellowing of young leaves and/or an interveinal chlorosis of older leaves; this is common when native plants of acid soils are moved to alkaline soils and can readily be induced in pot culture if 'hard' water is frequently given to sensitive species. Most horticulturists will be familiar

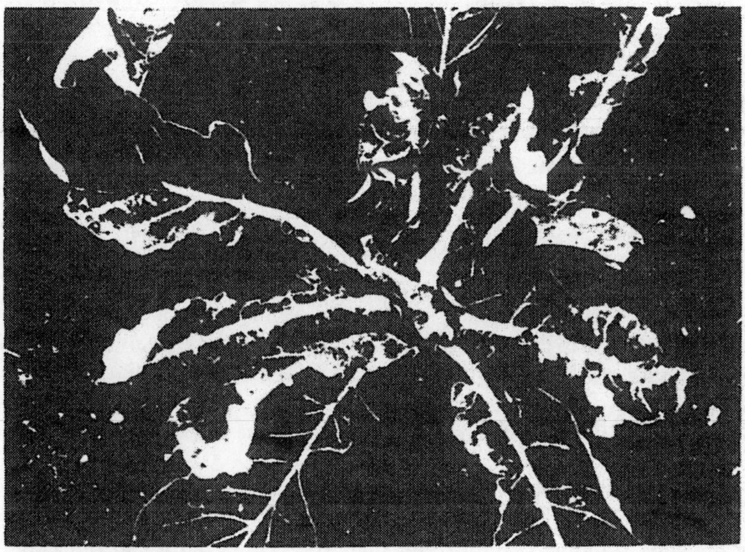

Figure 2.9 A shortage of the trace element molybdenum causes a condition in cauliflowers called 'whiptail'. The lamina of the leaf is often poorly developed causing a strap or whip-like appearance. The main growing point often aborts causing 'blindness' (NVRS copyright).

with the need to create acid soils by adding peat or sulphur for such plants as rhododendrons, azaleas and heathers. The common inorganic salts of iron present in most soils are relatively insoluble — and hence unavailable — if the *p*H is above 5.0. This is not true however of complexes between organic chelating agents and iron. Iron chelates are much more soluble and the iron is available even at high *p*H. One of the most commonly used chelating agents is ethylene-diamine-tetraacetic acid, which is fortunately abbreviated to EDTA. Zinc and copper are also better supplied as chelates as their inorganic salts are also relatively insoluble.

Water from the soil contains the essential elements of hydrogen and oxygen, the latter being obtained also from the air along with carbon as carbon dioxide. There are sixteen elements that are generally accepted, using Arnon and Stout's criteria, as essential for plant growth. These elements are used to synthesise the complex chemicals of plants. Each element has its part to play — some prominent like carbon, oxygen and hydrogen that give sugars; nitrogen being part of proteins; and phosphorus forming salts which are energy carriers. Calcium pectates have the job of cementing cell walls together and magnesium is part of the chlorophyll molecule. The studies of formation and function of sugars, proteins and other important biological substances are an important part of plant physiology and biochemistry, but their relevance to horticulture is limited. How is this chemical synthesis manifest as growth?

SOURCES AND SINKS

Simple inorganic nutrients are taken in by the roots and moved up the xylem to the sites of growth, there they and the sugars formed by photosynthesis are transformed into the substance of new cells. The energy-rich chemicals for growth are transported from the chloroplasts through the phloem. These energy-rich chemicals, such as sugars, are large soluble molecules not dissociated into ions. They have to travel up and down the plant, as well as laterally. Characteristically they travel from where they are produced (a source) to where they are used (a sink). The source is usually green tissue, but the sink may be the growing tip of a shoot, a developing flower or a root several feet below the ground. There is typically a high concentration of sugars at the source, whereas at the sinks the sugar concentration is a good deal lower.

The movement of these sugars is by mass flow of the whole solution (in diffusion the solvent remains stationary). In 1930 E. Münch described in his book a model system to account for this mass flow in the phloem, and this is still the basis of our understanding of the process. Münch showed that if two reservoirs, made of semi-permeable skins and containing sugar solutions of different concentrations, were connected by a tube (as shown in figure 2.10), then there was a mass flow from stronger solution (the source) to weaker solution (the sink), provided the two reservoirs

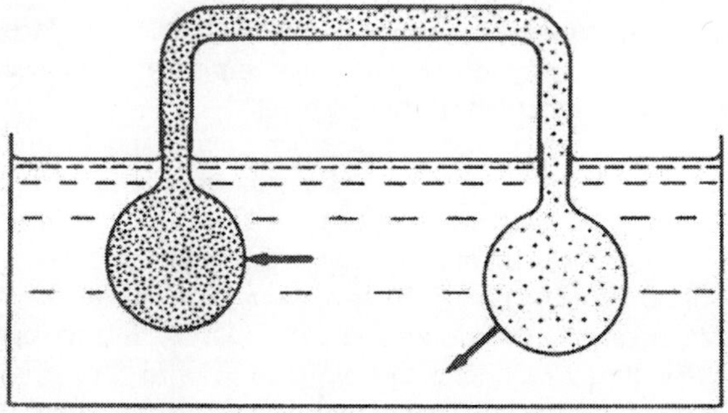

Figure 2.10 Münch's model of phloem flow (for explanation see text).

were both held in an even weaker solution. In figure 2.10 the same external solution bathes both reservoirs, but there would still be a flow along the tube if two separate external containers were used.

If we assume that the external solution is water, then the sugar solutions inside the membranes will have a lower chemical potential than the water, which will therefore pass through the membranes into the reservoirs. As we have seen earlier, the rate of this water movement becomes greater as the water potential increases. The higher the sugar concentration, the lower is the chemical potential and the greater is the water potential; consequently the flow of water into the source reservoir will be greater. Pressure will therefore build up in the source reservoir more quickly than in the sink reservoir. The pressures will be equalised by a flow of solution from source to sink, increasing the pressure in the sink reservoir until its chemical potential may even exceed that of the external water. When this happens, water will pass out of the sink; in effect there will be a flow of sugar solution from the source to the sink with water being returned to the source. The model does not require that water should move out of the sink. If the pressure in the sink were kept low by the reservoir expanding as the water entered, then flow along the tube would still occur.

Once the concentrations of sugars in both reservoirs are equalised, flow (in Münch's model system) would cease. Flow will continue however if sugars are being fed in at the source and used at the sink, so that the concentration is always being lowered.

This model system does not seem very like a living plant, but Münch realised that photosynthesising cells were the source in his model and the sieve tubes were the tube connecting it to the sink, which was any area of the plant using energy. Münch thought of the external weak solution bathing both source and sink as the water or weak solution in and between the cells surrounding the sieve tubes. Anatomists have examined the tissue surrounding sieve tubes and concede the possibility of Münch's idea. Whatever difficulties physiologists and anatomists find in accepting the theory, no other has emerged to take its place.

If we accept Münch's model of pressure flow, then all solutes within a tube ought to move at the same rate. But recent studies with radioactive tracers have indicated that this is not so. This anomaly has been explained without amending the theory by suggesting that diffusive mass flow may occur in the cytoplasm. Diffusive mass flow is like the carrying of substances at different rates along filter paper during chromatography.

It is difficult to reconcile this suggestion with the observed rates of movement of substances in phloem (20 to 100 cm per hour). This movement is achieved in spite of the presence of sieve plates and the cytoplasm and slime which fill the space. It has recently been claimed that tubules or strands can be seen connecting the pores of one sieve plate to those of the next and that material moves along these strands. It has been suggested that this structure might in some way aid flow but precisely how it does so is not clear.

Münch's model indicates that the sieve tube plays no active part in transport, but many experiments have shown that transport ceases if phloem is killed, suggesting that an active energy-requiring process is involved. On the other hand, transport may cease because the contents of the dead sieve tubes coagulate and block the lumens. There is also some evidence that metabolic inhibitors will slow down phloem transport — again suggesting that Münch's model is wrong. This observation is not easily explained away but it has been suggested that what is inhibited is the maintenance of the sieve tubes in a suitable condition for flow to occur; the flow is physical but the apparatus in which the flow occurs is living and requires maintaining.

It seems very probable that Münch's model is an over-simplification of the true state of affairs during movement in the phloem, but is probably on the right lines.

There is now little doubt that plant growth regulators (table 2.1) play an important role in determining the pulling power of sinks and that the transfer of sugars from the sieve tubes to the sinks is an energy-dependent process regulated by hormones.

DIRECTIONAL TRANSPORT

Some particularly mobile ions like potassium have been shown to circulate around the plant, going up the xylem and down the phloem. There is some evidence that larger molecules of sugars and some herbicides can also circulate around the plant in this way. Some substances, however, only move, or predominantly move, in one direction. This is termed *polar transport* and is important in the movement of plant hormones — and in particular, auxin. This hormone is synthesised in the young leaves and the apex of a shoot, and moves down towards the roots. This is not an effect of gravity because movement away from the apex occurs even if the shoot is inverted or horizontal. This polarity of movement is strongest nearer the shoot apex and weakest in the roots, where two-way transport has been observed.

Again the mechanism of this transport is not understood but it only occurs in living cells, is slower than would be expected for movement in vascular tissue but faster than diffusion would allow, and the maintenance of a capacity for polar transport somehow depends on a continuous supply of auxin. What is this auxin which seems to have its own method of moving through the plant? Natural auxin is a growth regulator, one of many such chemicals which occur in plants. The general nature and types of these plant hormones will now be discussed.

HORMONES

Plant hormones are chemical messengers that play a major role in organising growth and development. They are present in plant tissues in very low concentrations but their effect is often immense and dramatic. The presence of hormones was suggested by observation of plant growth long before experiments at the turn of the century eventually proved their existence. It was not until 1934 that the first plant hormone was prepared in a pure state from urine, and a year later before it was shown to occur naturally in plants. Many sophisticated and specialised techniques have since been developed to extract and purify plant hormones. These techniques now enable the quantitative and qualitative changes in the hormone content of plants to be related to changes in the pattern of growth. Examples of this will be given as we follow the growth of the plant throughout this book.

These natural growth regulators are classified into four major groups according to the response they produce. There are auxins, gibberellins, cytokinins and inhibitors. (Ethylene is a naturally occurring growth regulator but it cannot be readily fitted into these groups.) There are also a few synthetic growth regulators which have hormone-like action on plants but have not been found naturally. Among these are some herbicides which at very low concentration do not kill, but regulate growth. Table 2.1 classifies and summarises the effects of some growth regulators.

Hormones play a part in determining whether a plant is dwarf or tall; whether its leaves age and fall off; whether lateral buds develop; and so on. They affect all phases of plant growth and we have already seen that they can affect germination, in particular by overcoming seed dormancy but we shall deter further consideration of their action to later chapters.

QUANTITATIVE ASPECTS OF SEEDLING GROWTH

As the seed germinates and the plant becomes established, the root system branches and extends while the shoot forms leaves and the stem elongates.

Table 2.1

Plant growth regulators (after Thomas, 1971)

	Group	*Example*	*Effects*
	Auxins	Indoleacetic acid Naphthaleneacetic acid 2,4-Dichlorophen-oxyacetic acid	Rooting stimulation and fruit set stimulation Weed control
Naturally occurring and related synthetic regulators	Gibberellins	Gibberellic acid	Flower promotion Seed germination Stimulation of vegeta-tive growth
	Cytokinins	Zeatin Benzyladenine Kinetin	Bud development and inhibition of ageing
	Inhibitors	Abscisic acid	Induction of dormancy Promoting ageing Defoliation Stomatal control
	Growth retardants	Chlormequat Daminozide Phosphon	Reducing vegetative growth Stimulating fruit production
		Maleic hydrazide	Preventing dormancy-break
Other synthetic regulators	Herbicides	Simazine Carbamates	Various effects at low concentration
	Defoliants	Ethophon (Ethylene)	Inducing leaf drop Effects on flowering
	Disbudders	Methyl esters of fatty acids	Bud destruction

If we wish to study how different temperatures affect this growth, we must have a way of measuring growth.

At its simplest this might mean measuring the height of the shoot after keeping the plants at different temperatures for a month. This would not tell us very much because we would have no idea of how quickly temperature affected height; or whether it affected height by increasing internode length or by causing more internodes to elongate in the time. We would only have to count the leaves to settle this point but then we might ask if the length of internodes present before the temperature treatment began was affected as much as the length of those formed during the temperature treatment. The way we measure plant growth depends on the questions we want to answer. If we want to know whether bigger shoots obtained at higher temperatures meant there were smaller roots there is no alternative to making some measurement of root size.

Needless to say plant physiologists have developed what might be termed a standard questionnaire, so that when they wish to compare plants grown in different ways they would as a matter of course make certain measurements. Others would be undertaken to provide answers to specific questions. Although it is rarely put in this form, the most usual question is "What differences are there in the rate of fixation of energy between these sets of plants?". Growth is the net fixation of light energy and we can determine the energy of a plant by measuring the heat produced when it is burned (its energy value). The principles of the apparatus used are shown in figure 2.11. Dieticians use similar equipment to determine

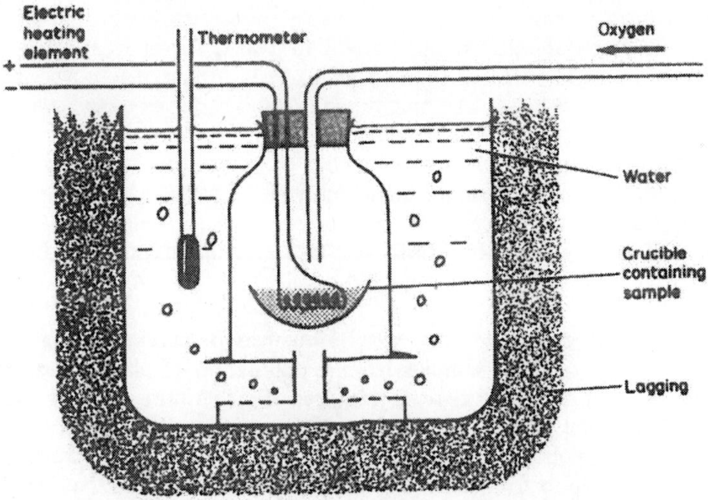

Figure 2.11 Diagrammatic section of a bomb calorimeter such as is used to determine the calorific value (energy content) of plant samples.

the energy value of foods and tell us that cabbages give 2.5 kJ g^{-1}, potatoes 3.9 kJ g^{-1} and tomatoes 0.7 kJ g^{-1}. Thus per unit of *fresh weight* these plant parts do not have the same energy value. A plant species is assumed however to have constant energy value per unit of *dry weight*. This assumption has become so generally accepted that many growth physiologists seem to have ceased to realise that it is an assumption and – at best – an approximation. Table 2.2 gives the energy values for the dry matter of the leaves and tap roots of red beet and carrots. Although both give similar values for leaves

Table 2.2

The calorific values of the dry matter of red beet and carrot leaves and tap roots, given as $kJ\ g^{-1}$ of dry matter

	Leaves	*Tap roots*
Carrots	14.4	18.6
Red beet	14.2	18.5

and tap roots, the tap roots have a much higher energy value than the leaves. Young plants are mostly leaf whereas old plants are mostly tap root. Clearly, if we wish to compare the energy fixed by young and old plants their dry weights would not give a very accurate indication of this energy.

The main reason for not measuring energy values is that it takes a long time. It is much easier to determine dry weight by simply putting suitably cut plant material into an oven at 98 to 100°C for at least 24 hours, cooking in a desiccator and then weighing. It is usual to dry to a constant weight so after an initial weighing the material is returned to the oven and reweighed after 12 to 24 hours. This process is repeated until a constant weight is obtained so demonstrating that complete drying has occurred. Dry weights are more accurate than fresh weights because plants that have wilted (either in the sun or as a result of being cut for weighing) have lost weight through water loss. Fresh weights are so inaccurate that they are rarely used unless measuring the yield of plant produce which is normally sold fresh.

Each measurement of dry weight involves destroying plants, so if growth is being studied samples from a population of plants have to be destroyed each time a weight is required. Furthermore, if we want to measure the total growth of the plant, the roots as well as the shoot must be dried and weighed. Very often physiologists find it too time consuming to recover the roots from the soil and so they further assume that ignoring the roots will not affect any conclusions they wish to draw. Many research papers ignore root systems so completely that the assumption seems to have become an axiom.

If a series of dry weights of seedlings is made by taking samples at intervals, a graph can be drawn from the data. Figure 2.12a shows that an upwards curve is produced. If the same weight increase had been made each day the line would have been straight and not curved. The curve tells us that an increasing dry weight gain was recorded for each day.

If the graph were a straight line, then growth would be like earning simple interest on an investment: the interest we earn is not added to the invested amount so that it too can earn interest. This only happens if an investment is subjected to compound interest; here the interest is continuously being added to the amount invested so that, although the rate of interest remains constant, the amount we receive each year increases.

Investment-conscious horticulturists who know about interest rates can see that increase in dry weight is like compound rather than simple interest. What rate of interest are we getting? This is difficult to see or to calculate either from the curve in figure 2.12a or from the data but if we plot the logarithms of the dry weights against time, the result is a straight line (figure 2.12b). The slope of this straight line is our interest rate and as long as it remains straight this rate is constant.

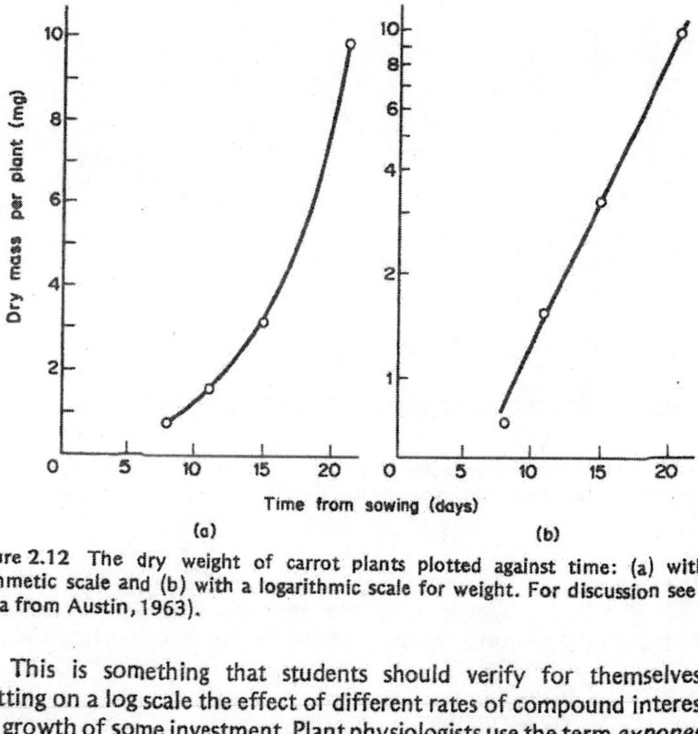

Figure 2.12 The dry weight of carrot plants plotted against time: (a) with an arithmetic scale and (b) with a logarithmic scale for weight. For discussion see text (data from Austin, 1963).

This is something that students should verify for themselves by plotting on a log scale the effect of different rates of compound interest on the growth of some investment. Plant physiologists use the term *exponential growth* for this compound interest. It is characteristic of the early growth of an individual seedling or seedling population, and the dry weight we plot is either the mean weight of a single plant or the weight per unit area of a young crop.

Relatively simple measurements of some plants demonstrate exponential growth and even the effects of environment on the rate of compound interest (termed the *relative growth rate*). Figure 2.13 illustrates some Japanese data on the growth of the pond weed *Lemna minor* (the Common Duckweed). Counting the fronds of this plant is a reasonably

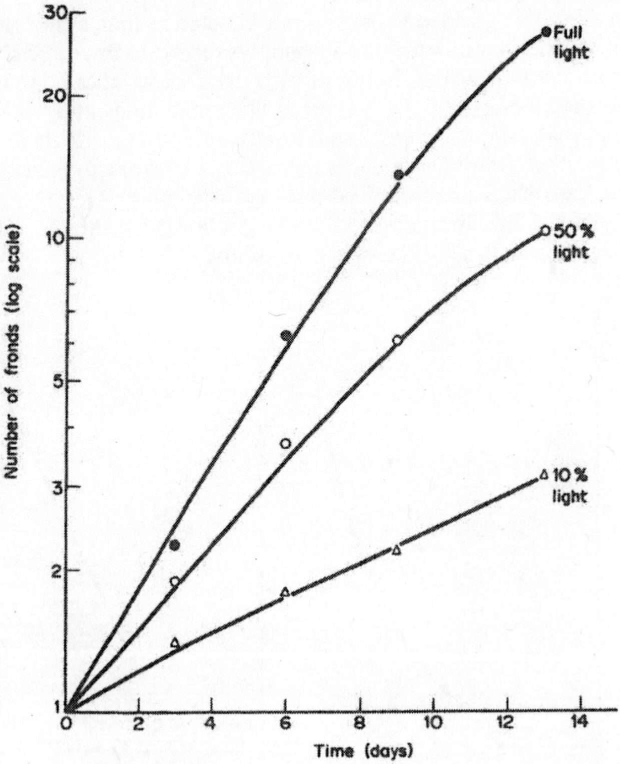

Figure 2.13 Increase in the number of fronds of *Lemna minor* (duckweed) at three light intensities (data from Ikusima and Kira, 1958).

accurate measurement of size, making this plant one of the classical materials for growth experiments. Over the first 9 days the number of fronds increases exponentially but the rate of increase is greatest at the highest light intensity. The relative growth rates are directly proportional to the gradient of the line. We can express this as the growth made in a set number of days divided by the number of days.

If we take the increase in frond number during 9 days, then in the highest light intensity the rise (measured on the graph paper I used to prepare figure 2.13) was 14.3 cm and the 9 days was equivalent to 9 cm.

Thus the gradient was: 14.3/9 = 1.59
Similarly at 50 per cent light 10.1/9 = 1.12
at 10 per cent light 4.2/9 = 0.52

Thus, the relative growth rate in full light was roughly half as much again as that at half light and three times that at 10 per cent of full light.

To put the above relative growth rates into absolute terms, we have to take into account the scales of the graph. One cycle (from 1 to 10) on the log scale was 12.5 cm. This distance is equivalent to $\log_{10}e$ (where e is the base for natural logarithms); hence our scale for the fronds was 12.5 cm = 2.303. Our scale for time was 1 cm = 1 day so no correction is necessary. Thus, to bring the relative growth rates to absolute terms they each have to be multiplied by 2.303/12.5. This gives values of approximately 0.3, 0.2 and 0.1 fronds per frond per day, respectively, for the highest to lowest light intensities.

This may seem rather complex but a little practice on compound interest rates will give the familiarity which will make for understanding. The relative growth rates can be calculated with tables of natural logarithms which are logarithms to the base e (written $\log_e$) rather than to the more usual base of 10. The equation used is

$$\text{relative growth rate } (r) = (\log_e W_2 - \log_e W_1)/(t_2 - t_1)$$

where W_1 is the weight (or some other measurement) at time t_1 and W_2 the weight (or whatever) at time t_2. If we take an example from a graph of compound interest at 10 per cent per annum, we shall see that at t_1 (zero on the graph) we invested 100 units (W_1) and this became 201 units after 7 years (t_2). So

$$\begin{aligned}\text{relative growth rate} &= \log_e 201 - \log_e 100/7 - 0\\ &= 5.3 - 4.6/7 = 0.1\end{aligned}$$

Therefore, we would get 0.1 units per unit per year or 10 units per 100 units or 10 per cent.

The equations can be altered; for example, given the relative growth rate we can calculate how long it would take for one frond of *Lemna minor* to cover one hectare surface of a pond. For the sake of this exercise, we shall assume that three fronds occupy one sq. cm and the relative growth rate obtained in full light by the Japanese workers. 1 hectare = 10 000 sq. metres = 100 000 000 sq. cm = 300 000 000 fronds.
Another form of the relative growth rate equation is

$$\log_e W_2 = \log_e W_1 + r(t_2 - t_1)$$

W_2 is 300 000 000 fronds, W_1 is one frond and $(t_2 - t_1)$ is what we wish to find. If t_1 is time zero, then $(t_2 - t_1) = t_2$.
$\log_e 1$ is also zero, so we have

$$\begin{aligned}\log_e 300\,000\,000 &= r\,t_2\\ \therefore 19.52 &= 0.3\,t_2\\ \therefore t_2 &= 19.52/0.3 = 65.6 \text{ days}\end{aligned}$$

The graph shows that only about 2 days are needed to double the number of fronds and it would not take long for all available fresh water surfaces to be covered with *Lemna minor*.

Clearly neither *Lemna* nor our seedlings can continue to grow exponentially for more than a short period. Growth must be regulated in some way and this is explained in the next chapter.

References

ARNON, D. I., and STOUT, P. R. (1939). The essentiality of certain elements in minute quantity for plants with special reference to copper. *Pl. Physiol., Lancaster*, 14, 371.

AUSTIN, R. B. (1963). A study of the growth and yield of carrots in a long-term manurial experiment. *J. hort. Sci.*, 38, 264–76.

BALEGH, S. E., and BIDDULPH, O. (1970). The photosynthetic action spectrum of the bean plant. *Pl. Physiol., Lancaster*, 46, 1–5.

HEATH, O. V. S. (1969). *The Physiological Aspects of Photosynthesis*. Heinemann, London.

IKUSIMA, I., and KIRA, T. (1958). Effect of light intensity and concentration of culture solution on the frond multiplication of *Lemna minor* L. *Seiro-seitai (Physiology and Ecology)*, 8 (1), 50–60.

MÜNCH, E. (1930). *Die Stoffbewegungen in der Pflanze*. Fischer, Jena.

SACHS, JULIUS (1875). *Textbook of Botany; Morphological and Physiological*. Clarendon Press, Oxford.

THOMAS, T. H. (1971). Plant growth regulators and crop production. *N.A.A.S. q. Rev.*, 89, 33–42.

Further Reading

de WIT, C. T. (1965). *Photosynthesis of leaf canopies*. Pudoc Agricultural Research Report No. 663.

ELECTRICITY COUNCIL (1980). *Growing Rooms*. Farm-electric Centre, Stoneleigh, Warwickshire.

ELECTRICITY COUNCIL (1981). *Lighting in Greenhouses*. Farm-electric Centre, Stoneleigh, Warwickshire.

ELECTRICITY COUNCIL (1982). *Electricity in Horticulture*. Farm-electric Centre, Stoneleigh, Warwickshire.

GAASTRA, P. (1959). Photosynthesis of crop plants as influenced by light, carbon dioxide, temperatures and stomatal diffusion resistance. *Meded. LandbHoogesch. Wageningen*, 59 (13) 1–68.

PHILIPS (1982). *Artificial Lighting in Horticulture*. N.V. Philips, Eindhoven, The Netherlands.

THOMAS, T. H. (1982). *Plant Growth Regulator Potential and Practice*. BCPC Publications, London.

WHATLEY, J. M., and WHATLEY, F. R. (1980). *Light and Plant Life*. Edward Arnold, London.

CHAPTER 3

THE VEGETATIVE PLANT

We have seen in the previous chapter that the early growth of seedlings can be likened to earning compound interest and that this is termed *exponential growth*. We also saw the prodigious growth of one frond of duckweed (*Lemna minor*) if it maintains this exponential pattern. It is amusing to use the equations given to calculate how large an area one frond could cover in a year, or even how long it would take to cover the earth's surface. Science fiction thrives on these unreal situations. What factors normally constrain plant growth?

Physiologists study the growth of plants to find what these constraints are and how they operate. In such studies, as we saw in the last chapter, the dry weight of the plant is often a favoured measurement but there are others. There has grown up something of a convention about what should be measured and how the data should be summarised and examined. This collection of growth measurements and its subsequent examination and interpretation is called growth analysis.

GROWTH ANALYSIS

The basis of growth analysis is the sequential measurement of growth and the basic measurement of growth is the dry weight. Because this is destructive, different plants must represent at each sampling the populations being studied. It is most important, therefore, that the sample plants should be representative of the population and proper statistical procedures must be followed. Apart from dry weights, often of individual plant parts (for example, leaves, stems, roots), it is usual for leaf areas to be measured.

Measurements or estimates of leaf area have been made in various ways. Some of the modern methods either use television cameras as scanning devices, or a graphics tablet and light pen, or estimate leaf area from its impedence to a controlled airflow. The traditional equipment for measuring leaf area is a planimeter; a pointer is steered round the edge of a leaf until the starting point is again reached and the area of the leaf is shown on a dial. The leaves must be trapped below glass to give a convenient 'table' for the planimeter and great care has to be taken to ensure that the mechanism does not skid. Each leaf has to be measured separately and more than once to ensure accuracy. It is impossible to assess the area of a very divided leaf, such as that of a carrot, by using a planimeter, but fortunately there are now more reliable and more accurate methods available.

The principle involved in some of these methods was first proposed by some Japanese workers, although I and my colleagues at Wellesbourne independently developed the 'dot' method. The basic apparatus is simple. If placed over a graph paper divided into 1 cm squares, a piece of glass or clear plastic can easily be marked with a very small dot at the corners of each square. This gives a clear plate with tiny dots spaced at regular 1 cm intervals. If this is then placed over a leaf and the number of dots over the leaf ('hits') are counted, then this count is the area of the leaf in square centimetres. It is essential that one uses small dots, assessing if the centre is over the leaf, and viewing must be at right angles to the glass to avoid parallex errors. This system will work even if the leaf is extremely divided or even filamentous. It can be proved to be mathematically sound and its accuracy can easily be demonstrated with shapes of known area. By spacing the dots at different intervals, other scales can be used. Like any other system based on sampling, it is important that a minimum number of hits is recorded to maintain accuracy. This can be done by using a sufficiently small scale or by calculating the mean of multiple readings. For example, if we wish to measure a leaf, about 2 sq. cm in area, we cannot expect an accurate answer if our dots are 1 cm apart — they would have to be 0.25 cm apart to get an accurate answer. Similar accuracy could be obtained with a 1 cm grid by taking sixteen counts, moving the grid slightly each time, and averaging the counts obtained.

Of course more than one leaf can be placed under the dotted glass at one time. If they overlap or are folded it is easy to count two for that dot. Alternatively, if the glass is fully loaded, it is sometimes quicker to count the 'misses' and subtract these from the total number of dots known to be on the glass. The total area of a leaf and the area affected by some discoloration are also easily assessed, even if the discoloration is itself a series of small dots. There would be little difficulty in assessing the areas of different colours on a *Coleus* leaf and comparing younger and older leaves on the same plant to see if they differed in this respect.

The dot method is sometimes too slow, because it often involves a considerable amount of counting. In these instances the expensive machines based on this principle become worthwhile. But even without them there are acceptable ways of reducing the work and the tedium. For example, it may be possible to establish the relationship between area and weight on a small sample and subsequently calculate leaf areas from leaf weights. This can be done using the dot method to determine the area or by using a cork borer to obtain fifty pieces of known area, which are then weighed. It is usual to use dry weights as this improves accuracy.

In many instances a good relationship can be established between one or two simple linear measurements of a leaf and its area. The precise nature of this relationship will, of course, depend on the shape of the leaf and the measurement taken, and will only apply to plants similar to the one used to establish the relationship. The establishment of such relationships is useful because the linear measurements can easily be made on leaves still attached to plants. Consequently this technique is frequently adopted in studies of the change of leaf area with time.

It is usual to regard the leaf area as the plan area of the leaf and not the surface area. That is to say only the area of one side is recorded and used in subsequent calculations designed to reveal facets of the growth of the plant or crop.

Leaf area and dry weight are measured at intervals and often plotted against time in simple graphs or – for reasons made apparent at the end of the previous chapter – the logarithm of the weights may be plotted against time. These simple measurements can also be used to calculate various parameters, which in turn may be plotted against time. For example, from the measurements of leaf area the area of leaf per unit area of ground on any one occasion can be calculated. This is termed the *leaf area index* and is abbreviated to LAI or L. If leaf area index is plotted against time the area under the curve is a measure of the area and duration of the leaves and is termed **leaf area duration** (LAD or D).

Net assimilation rate

The leaf area is measured because the leaves receive energy from the sun, and if we know their area and the change in plant weight over a limited period, we can calculate how effective these leaves have been. We will then know how much dry weight each unit area of leaf has assimilated in that time. We can then see if any treatments make the leaves more efficient or if the efficiency merely changes with time. This measure of efficiency is termed the *net assimilation rate* and is abbreviated to NAR or E. It is usual to calculate NAR using leaf areas but arguments can be made for using other measures of photosynthetic potential, such as leaf weight, leaf protein or nitrogen content. These all give different values for E and often different trends with time (see figure 3.1). To distinguish these values, E_A represents an estimate of NAR based on leaf area; E_W an estimate based on leaf weight, and so on.

The generally accepted equation for calculating net assimilation rates is

$$E = (W_2 - W_1/L_2 - L_1) \times (\log_e L_2 - \log_e L_1/t_2 - t_1) \qquad (3.1)$$

where W_1 is the total plant weight at time t_1, W_2 the weight at t_2 and L_1 and L_2 the leaf areas at time t_1 and t_2 respectively; L_1 and L_2 could also be the leaf weights or leaf nitrogen contents, as mentioned above.

Growth rate

If we wished to analyse the yield of a crop as opposed to that of a single plant we would be interested in the **crop growth rate** (C), which is the rate of increase in weight per unit of ground. As net assimilation rate (E) is the

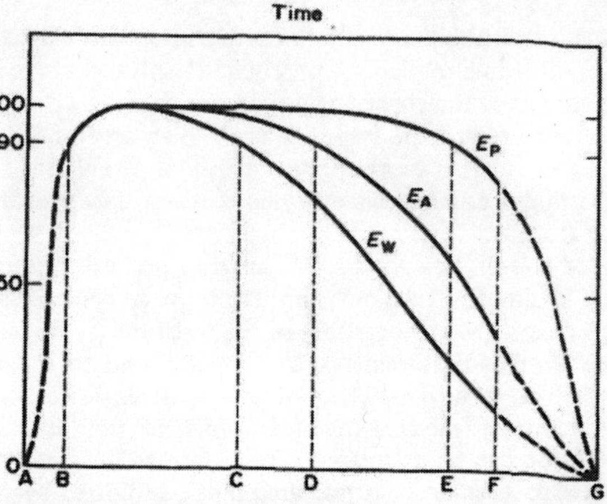

Figure 3.1 Diagram showing the time trends in net assimilation rate, calculated on the basis of leaf weight (E_W), leaf area (E_A) and leaf protein-nitrogen (E_P). The curves represent the expected time trends for the growth period of an annual plant in a constant environment and with nitrogen in relatively short supply. If the seedling phase (AB) and the senescent phase (FG) are disregarded then E_P is more constant than E_A or E_W (E, D and C represent the respective points for a 10 per cent decline). It is suggested that over the period BF, E should not show a decline in a constant environment and it is concluded that E_P is the best available measure of E (Williams, 1946).

rate of increase in weight per unit area of leaf and leaf area index (L) is the area of leaf per unit area of ground, it follows that

$$C = L \times E \tag{3.2}$$

The implications of this simple equation have filled the minds of crop physiologists for decades. At its face value the equation says that if we want to increase C we must increase either L or E. As E is the reflection of rather basic processes (photosynthesis and respiration), the rate of which might prove difficult to alter, effort was put into increasing L. It seemed self-evident that this would work because nitrogen fertiliser increased leaf (and hence L) and also gave higher yields. One could increase L at an early stage by having more plants and this often gave higher yields. Unfortunately, this thinking is now known to be too naive. Figure 3.2 is a diagrammatic representation of the effects of the number of plants per unit area on total weight per unit area on four occasions. At the earliest harvest (1) the weight per plant has not been affected by competition and there is a direct proportionality between weight per unit area and density. As time passes, competition will reduce the mean weight per plant at the higher densities to give a nearly constant yield per unit area over an increasingly wide range of high densities. At the time of the second harvest (2 in figure 3.2) and

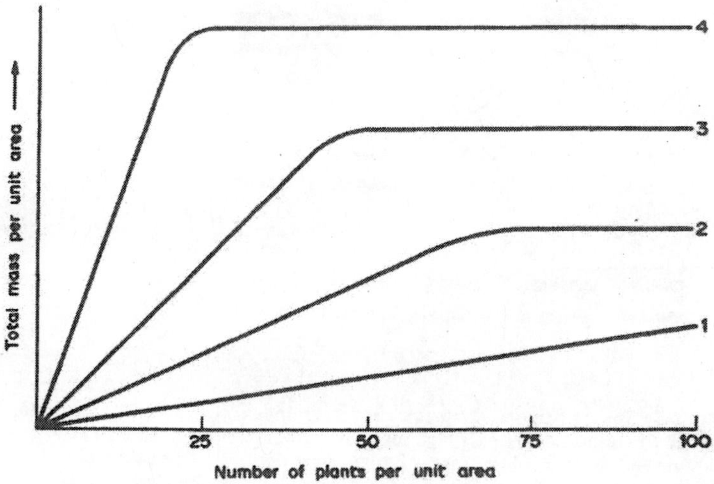

Figure 3.2 Diagrammatic representation of the relationship between plant population and total weight per unit area on four occasions (for discussion see text).

thereafter, the weights per unit area of plants at densities of 75 to 100 are the same at any one time. This implies that from the time of the second harvest onwards, plants growing at densities of 75 to 100 had the same relative growth rate, which was the lowest of any exhibited by plants at any density. This means that the relative growth rate of plants at lower densities was either the same as, or higher than, that of the plants at the highest density.

The simplest interpretation of this constancy of relative growth rates at high densities is that the canopies of leaves are similar over a wide range of high densities. This similarity of canopies is responsible for the similarity of relative growth rates (figure 3.3). This suggests that at any one time during crop growth there is a ceiling canopy that cannot be exceeded by increasing the plant density. If we are dealing with a competing crop, we cannot therefore increase *L* by increasing plant density and so increase *C*.

THE RESTRICTION OF GROWTH

The canopy clearly plays a part not only in providing the photosynthates for growth, but also in restricting growth. To begin to understand this we can return to the Japanese data on duckweed that provided the example of exponential growth at the end of chapter 2. If the initial exponential (compound interest) growth had been maintained, one frond would have grown to cover a hectare in about 65 days. As this does not happen, something reduced the growth rate. In figure 2.12 we can see that at the

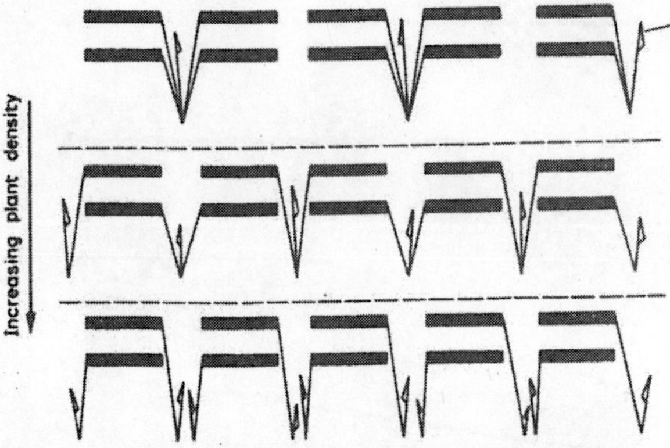

Leaf layer	P	R	Net assim of foliage
1	12	2	10
2	10	2	8
	22	4	18

Figure 3.3 The maximum attainable leaf area index (LAI) in a given time, and under given conditions by very high plant densities is termed the interim-ceiling LAI. It is constant over a wide range of high densities. For discussion see text. P = photosynthesis; R = respiration. All three canopies give the same net assimilation rate.

two higher light intensities the straight line bends downwards. The slope of this line is a measure of the relative growth rate and the more it tends to the horizontal, the lower is this rate. In this example the reduction in relative growth rate is not dramatic. The same Japanese workers also gave data for growth at different levels of nutrition (different concentrations of the nutrient in which the *Lemna* was growing). Figure 3.4 shows the increase in number of fronds with full nutrient solution and with a one-sixteenth strength solution. Even in the full nutrient solution there is some decline in the relative growth rate with time, but in the weaker nutrient this decline is rapid and the relative growth rate soon becomes very low (the line is almost horizontal).

This must be because there is less nutrient available. Even the fronds in the full nutrient solution would cease to grow once the nutrient had been depleted. The rate at which the fronds increased was not affected by the nutrient level for the first 3 days, when it was similar to the rate at the highest light intensity shown in figure 2.12. In this figure the fronds in 10 per cent light maintained exponential growth throughout the 13 days. They had full nutrient but the *rate* at which they could use it in growth was limited by light, eventually stabilising the number of fronds because all the light would be needed to maintain the fronds already produced. The simplified picture that emerges is that growth proceeds exponentially until the supply of one or more factors needed for growth becomes limiting. The concept of limiting factors and its implications have already been discussed in chapter 2. Even at reduced rates, growth continues until limiting factors prevent further growth (more strictly increase in dry weight).

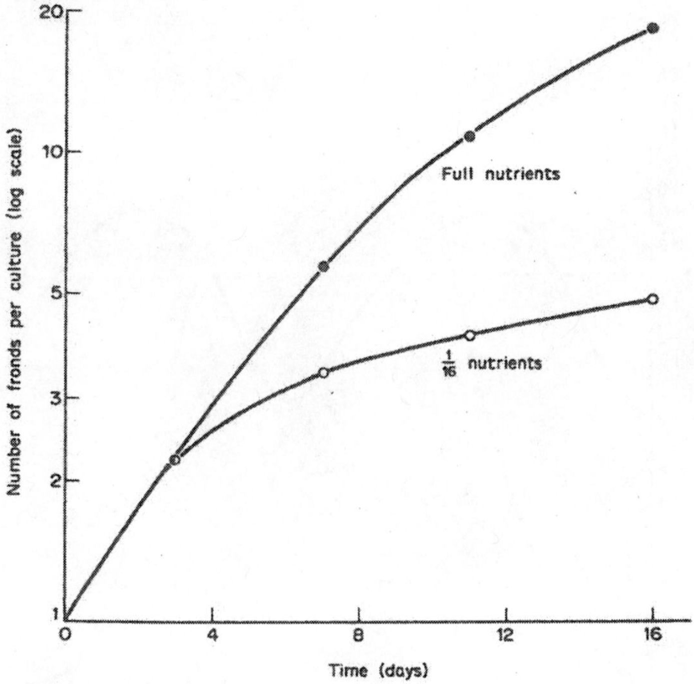

Figure 3.4 The increase in the frond number of duckweed (*Lemna minor*) at two nutrient levels (data from Ikusima and Kira, 1958).

Leaf area index takes no account of the arrangement of leaves. In crops grown in widely spaced rows the leaf area index is greater above the rows than it is between the rows (figure 3.5). This uneven distribution can mean that all the phases of leaf area index shown in figure 3.6 may be found within a crop. We also saw in chapter 2 that the angle of leaves could affect the growth due to a given leaf area index; vertically disposed leaves shade each other less than horizontal ones and are less likely to receive excessive light.

The canopy

The plant will cease to increase in dry weight when the rate of photosynthesis equals that of respiration. The plant may continue to grow by shedding old leaves and producing new ones. Some fairly distinct phases in this balance between input and output are best summarised diagrammatically as in figure 3.6.

Increasing yield is largely a matter of increasing the interim-ceiling leaf area index (figure 3.3). Put into agronomic terms, supplying extra water and nutrients or improving the temperature increases the amount of

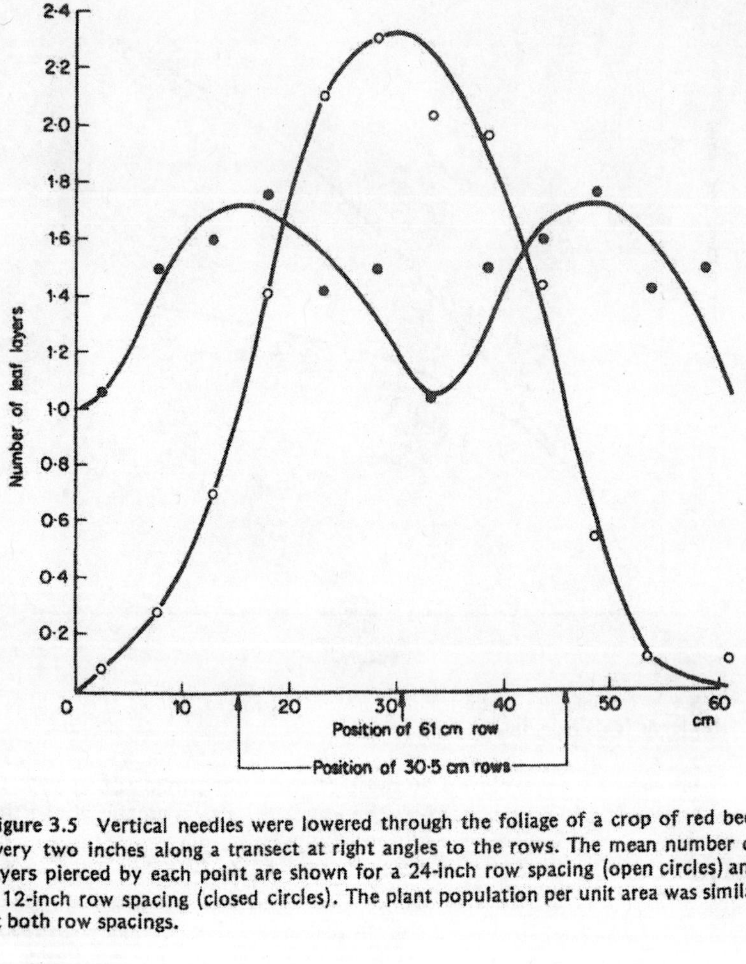

Figure 3.5 Vertical needles were lowered through the foliage of a crop of red beet every two inches along a transect at right angles to the rows. The mean number of layers pierced by each point are shown for a 24-inch row spacing (open circles) and a 12-inch row spacing (closed circles). The plant population per unit area was similar at both row spacings.

growth that can be achieved per unit area of ground in a given time. Simply having more leaf does not increase the rate of growth or the time available. Of course there are circumstances where having more leaf, in the form of extra plants, would increase yield but from studying figure 3.2 we can see that this would only be where the initial plant densities were too low to give maximum yield at the final harvest.

TIME AND GROWTH

In the preceding discussion growth rate has been expressed using days. Although legitimate and constant units of time, days are not constant units when considering plant growth. In a temperate climate, a summer

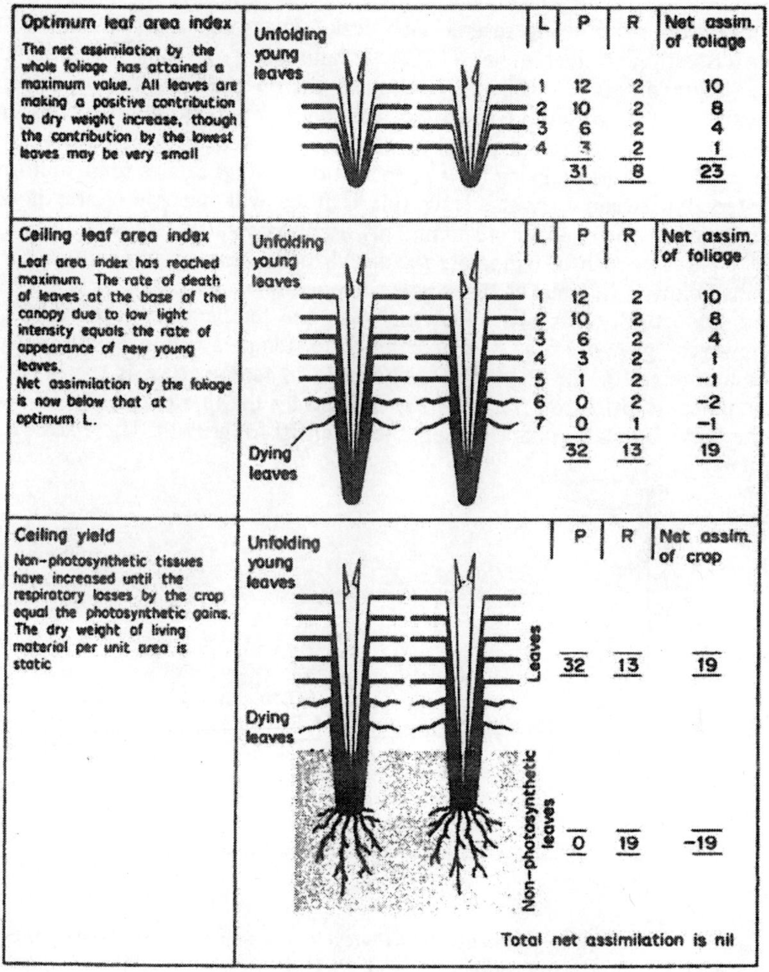

Figure 3.6 Diagram showing the status of photosynthesis (P) and respiration (R) of a crop at three different stages of crop and canopy development (reproduced from Donald, 1961).

day usually sees more plant growth than a winter day with lower temperatures and light intensities. If growth analysis is carried out on field plants, then changes in their growth rate or the relationship between NAR and LAI probably depend more on seasonal changes in the weather.

Some scale other than astronomical time might reveal more truthfully the effects on growth of treatments such as different fertiliser levels. The implications for using several such scales in growth analysis have been described by Nelder *et al.* (1960). Measurement of total radiation (or radiation of 400 to 700 nm) falling on a crop could reflect the energy available for growth. Growth in the field often closely depends on temperature and several practical time scales based on temperature have proved useful in knowing how to space out sowings to give some uniform interval between harvests. How this is done will become clearer if we understand one of the more commonly used temperature scales.

Plant growth in temperate regions virtually ceases if the temperature falls below 6°C. Thus at times when temperatures are below this, there is no opportunity for plant growth. When the temperature is above 6°C, however, growth rate is proportional to temperature over the range encountered in the field. The daily cycle in temperature is represented graphically in figure 3.7. The *area* enclosed by the line drawn at 6°C and the curve indicates quantitatively the potential for growth. These areas are

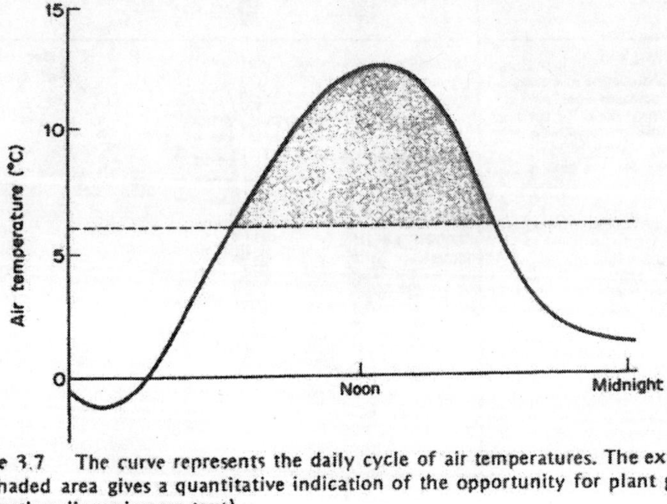

Figure 3.7 The curve represents the daily cycle of air temperatures. The extent of the shaded area gives a quantitative indication of the opportunity for plant growth (for further discussion see text).

in units of days and degrees (day-degrees). The Meteorological Office in London has issued tables which enable day-degrees to be estimated from the daily maximum and minimum air temperatures. Although a common practice, it is wrong to assume that an adequate calculation has been performed if 6°C is subtracted from the daily mean temperature.

The average day-degrees per week throughout a year reveal a great deal about the opportunities for growth. For example, figure 3.8 shows that at the National Vegetable Research Station (which is almost in the centre of England), a week in April provides only the same opportunity for growth as two days in July.

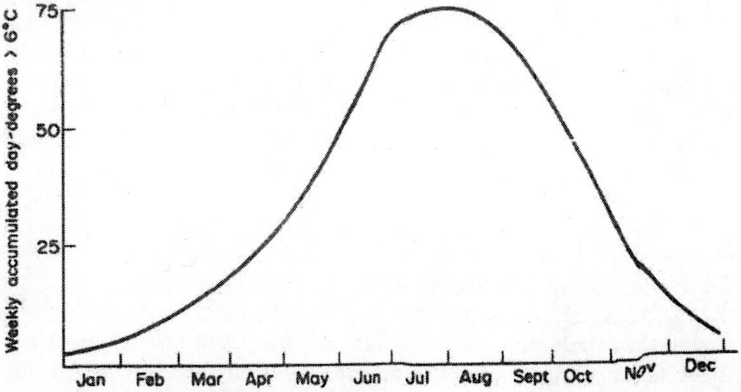

Figure 3.8 The average day-degrees (above 6°C) per week at the National Vegetable Research Station, Wellesbourne, Warwickshire, England (NVRS copyright).

This day-degree time scale is too simple; for example, it is assumed that any 1°C rise in temperature has the same effect on growth provided it is above 6°C. We know that too high temperatures prevent growth, so it would seem reasonable to set some arbitrary level (say 35°C), above which no growth occurs. Furthermore growth is probably neither increased nor decreased by a range of temperatures below this level, say between 30 and 35°C; any temperature between these limits should be reckoned as though it were 30°C. This juggling with limits can be useful but there is the danger that if one juggles enough the required answer will be obtained and that this answer will be a tribute to mathematical dexterity devoid of biological significance.

I have found it convenient to term ten Celsius day-degrees as a growth

Table 3.1

Growth units per week based on data for 1952–63, Wellesbourne, England

	Month											
Week No.	*J*	*F*	*M*	*A*	*M*	*J*	*J*	*A*	*S*	*O*	*N*	*D*
1	0	0	1	2	3	6	8	7	6	5	2	0
2	0	0	1	3	5	6	7	7	6	4	2	0
3	0	1	2	3	4	7	7	6	6	3	1	0
4	0	1	2	3	4	7	8	7	5	3	0	0
5	0	—	—	—	5	—	—	7	—	—	—	0
Totals	0	2	6	11	21	26	30	34	23	15	5	0
Running	0	2	8	19	40	66	96	130	153	168	173	173
Totals	173	171	165	154	133	107	77	43	20	5	0	0

unit. The main advantage of this growth unit scale is that a table of weekly mean values for a temperate site (table 3.1) contains only single figures, easily summated over periods representing the expected life of a crop. The Ontario Heat Unit, developed by Brown (1969), uses a lower base temperature during the night than during the day and has been particularly successful with sub-tropical crops such as maize and tomatoes.

Modelling growth

A knowledge of the structure of the canopy (as defined by leaf area, angles, and layers) can be used to calculate the light falling on each leaf. Knowing the relationship between light intensity and photosynthesis it is possible to calculate the photosynthetic input by each leaf for any light intensity. The plants' respiration can also be calculated and then subtracted from the photosynthetic input to obtain net input, which must then be apportioned to the different plant parts according to certain rules based on observation. We can add to the model the effects of carbon dioxide concentrations on photosynthetic rates, or the effects of temperature and nutrients on the apportioning of assimilates for new growth.

We use the computer to test the effect of new characteristics that may improve the yield of a plant. For example, we can model the outcome of changing the allotment of assimilates so that less goes to unwanted leaf and more goes to the bulb or root to be harvested. It might prove best to select from segregating lines at a seedling stage, when the required difference would be very apparent.

Unfortunately, our knowledge is so limited that it is unwise to place too much faith in prediction based on complex models. Modelling growth has indicated where new knowledge is required. The 'source–sink' control of photosynthesis is just one of the basic problems of plant physiology being studied. Simply stated, we would like to know whether photosynthetic rate is at all limited by the accumulation of unused products of photosynthesis. In other words, does the source (the leaf in light) or the sink (the use of photosynthates) control photosynthetic rate? The answer is apparently that sink control is possible, though source control is more likely. This subject is still controversial.

Computer modelling has stimulated the study of the physiology of plant growth. Relatively simple models have been used to predict the performance of crops in new areas and in the range of weather likely to be encountered. For example, Gray (1981) predicted that the first harvests of outdoor tomato crops in the UK at 53°N, 52°N and 51°N would occur on 12 August ± 6 days, 9 August ± 8 days and 4 August ± 4 days, respectively. The corresponding dates for a yield of 37.5 t/ha were 20 September ± 11 days, 13 September ± 11 days and 4 September ± 4 days. He also plotted areas of the country where yields of 37.5 t/ha could be expected for nine years out of ten.

COMPETITION BETWEEN PLANTS

Crop plants are seldom grown so widely spaced that they do not interfere with each other. We have already referred to the crop growth rate (page 55) and seen how plant numbers can affect yield at different harvest dates (figure 3.2). It was pointed out that to achieve the maximum total weight per unit area the plants must be competing; the bigger plants take more of the available resources, the smaller plants get less and many even die before the crop is harvested.

Weeds

One of the axioms of horticulture is that the removal of weeds is beneficial. By definition, a weed is any plant in the wrong place and it takes some of the resources that could have gone to the crop. Weeds can restrict growth and, even when removed early in crop growth, a detrimental effect can be still apparent at harvest. It is reasonable to suppose, however, that there is usually a period at the beginning of a crop's life when the presence of weeds will have no effect on the final yield. Similarly allowing weeds to develop when a crop is nearing harvest also has little or no effect on yield. Between these two periods, the crop is vulnerable to weeds. Nieto *et al.* (1968) designated this as the critical period for weed competition and presented the situation graphically as in figure 3.9.

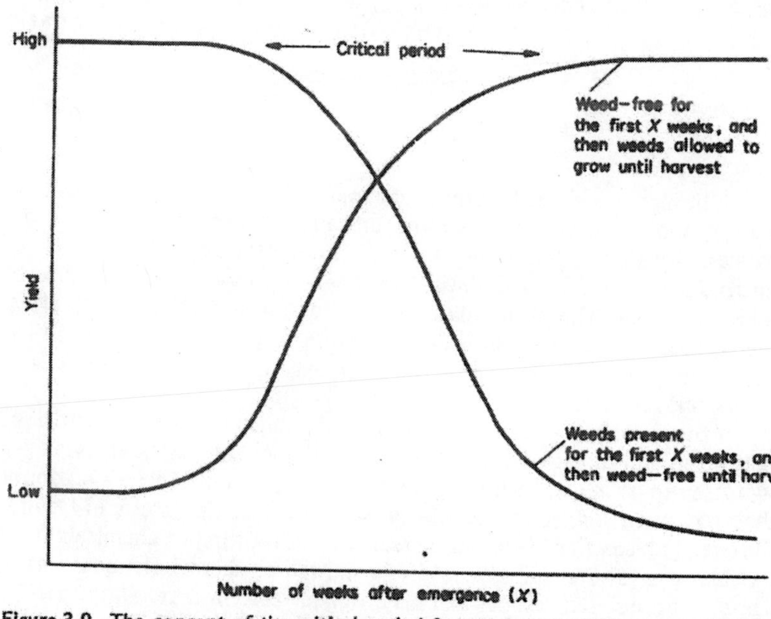

Figure 3.9 The concept of the critical period for weed competition as proposed by Nieto *et al.* (1968).

Most weeds seem to detract from the yield of crops according to their size (or more strictly their weight). The weeds and the crop plants are parts of a competing system. Horticulturists are aware that ground-cover plants will keep down weeds and the canopy of some annual crops can minimise the effect of weeds. A simple experiment that I carried out several years ago illustrates this point.

A crop of red beet was mechanically hoed soon after emergence and this left a band of weeds coincident with the crop rows. On other plots a similar band of weeds was left *between* the rows. No other weed was allowed to grow and on some other plots all weeds were removed as soon as they were big enough to grasp. The results are given in table 3.2. The weeds within the row, which were seemingly in a better position to suppress crop growth, grew less and suppressed the crop less than the weeds left between the crop rows. These latter weeds were not affected at first by the crop; they flourished and established a claim to part of the ground that the crop would have been able to utilise in their absence.

Table 3.2

The effects of position of weeds on the yield of red beet (*Beta vulgaris*), given as kg per 15 m of row

ITEM	*TREATMENTS*			
	Weed-free	*Weeds in crop rows*	*Weed between crop rows*	*S.E. of difference between two means*
Tops	41.9	38.6	31.6	±2.9
Tap roots	43.6	36.0	30.9	±1.7
Weed	none	2.6	4.4	±1.0

Although most weeds seem to compete with crops in relation to their weight, some plants species exude substances which are toxic to other species. This is comparable to the production by one species of micro-organism of an antibiotic which will kill some other species of micro-organism. The term **kolines** is used for the substances which affect the competitive relationship between some species of higher plants. These kolines may be volatile essential oils from leaves, or lactones, peptides or amino acids from roots.

Many laboratory experiments have demonstrated the potential of kolines in limiting growth, but these do not prove that kolines play a significant part in competition under normal conditions. There is no doubt that the weed *Camelina alysium*, which occurs in flax crops in Central Europe, reduces flax yields as a result of the washing of chemicals from its leaves into the soil by rain. Furthermore, there have been many demonstrations that kolines can play a major part in determining species distribution in arid areas.

Kolines are of very limited horticultural significance, although they could explain why some choice plant will not grow where required. If this preferred but hostile site happens to be near a plant of another species with aromatic foliage, then you may well be witnessing chemical warfare!

Crop spacing

The horticultural management of competition between crop plants reached its peak in the heyday of French gardening. High production was maintained by inter-cropping so that no one crop plant was allowed more room than it needed at any time. These systems of production have been replaced nowadays by single cropping, which lends itself to mechanisation. The horticulturist is as keen as ever to achieve the maximum production possible. Apart from any commercial incentive, there is an element of thriftiness in horticulture that deplores any under-utilisation of available resources. The spacing of the individual plants of a crop determines more than any other single factor the resources available to each plant and whether or not the total resources are fully utilised.

Patterns of arrangement

Spacing is not simply the number of plants per unit area, for plants can be arranged in many different ways. Jethro Tull and Turnip Townsend introduced the growing of crops in rows to British agriculture in the mid 18th century. When horse-drawn hoes were used, the rows had to be far enough apart to allow a horse to pass, and the advent of tractors did not immediately reduce row spacings. The increasing use of herbicides has made inter-row cultivation unnecessary, and for the first time in two hundred years we are free to arrange crops for maximum use of the land. With widely spaced rows, neighbouring plants will compete with each other long before they can use the inter-row space. This can be represented diagrammatically by allocating to each plant a circle of influence, as in figure 3.10; where the circles overlap, there is competition. Regular spacing — where the within-row and between-row spacings are equal — can delay the onset of competition and make better use of the ground.

If a constant row spacing is used in experiments involving different plant populations per unit area, then two factors are being varied simultaneously and their effects on plant growth cannot be separated. The two factors are plant population and rectangularity (the ratio of the between-row spacing to the within- row spacings). For example, if two similar crops are grown in rows 50 cm apart, one with plants 10 cm apart and the other with twice as many plants 5 cm apart, then the rectangularities are 5:1 and 10:1, respectively.

Rectangularity ratios are a convenient way of describing the distribution pattern of row crops, but these ratios are of little significance unless the plant population is also stated. Perhaps this can be readily seen if we imagine some absurdly low plant population (like one plant per

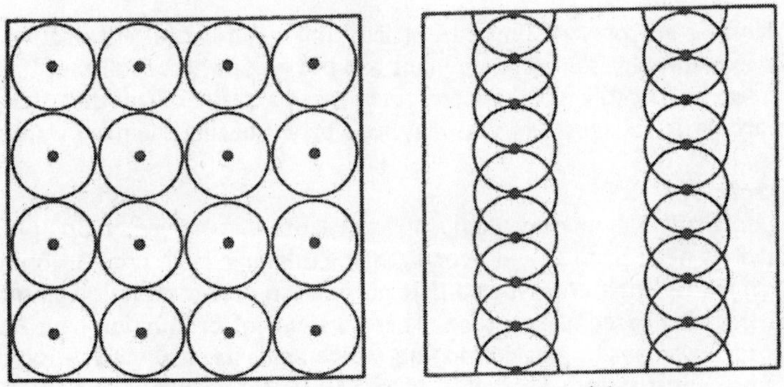

Figure 3.10 Diagrammatic representation of the effects of pattern of plant arrangement on competition. Each plant is represented by a dot and its area of influence by a circle. With the square arrangement (a), the circles do not overlap and most of the total area is being utilised. With arrangement (b), the overlapping of the circles indicates severe competition between plants and in addition there is under-utilisation of the available space.

hectare) and realise that a very wide range of rectangularities at this density would not affect yield very much. At high plant population rectangularity becomes of great significance but it still may not represent a meaningful scale for studying the effects of pattern of plant arrangement. This appears to be so because crops grown at high population are not usually spaced regularly and the within-row irregularity in spacing is significant. This is not accounted for by the rectangularity ratio, because this expression is obtained by using the mean within-row spacing.

Other methods of defining the space allocated to each plant of a population in terms of area and shape have been – and are being – developed. One proposal is that each plant should be thought of as being able to utilise a polygon delineated by lines drawn at right angles to the mid-points of lines joining the plant to its immediate neighbours (figure 3.11). On this basis, not only is the area of the polygon important but also the off-centredness of the plant within its polygon.

This system of defining distribution does not seem to be fully satisfactory, although it does focus attention on the fact that irregular distribution gives a wide range of areas and opportunities to plants in a population and that a wide range of plant sizes must be expected.

Other factors in a competing system also cause size variation; some variation may be genetic but competition probably affects its expression. Such factors as time of emergence, size of cotyledons and rate of seedling growth may all be under genetic control but early emergence and a large size will give a plant an advantage over its competing neighbours that emerge later and grow more slowly. Once this advantage is established, the

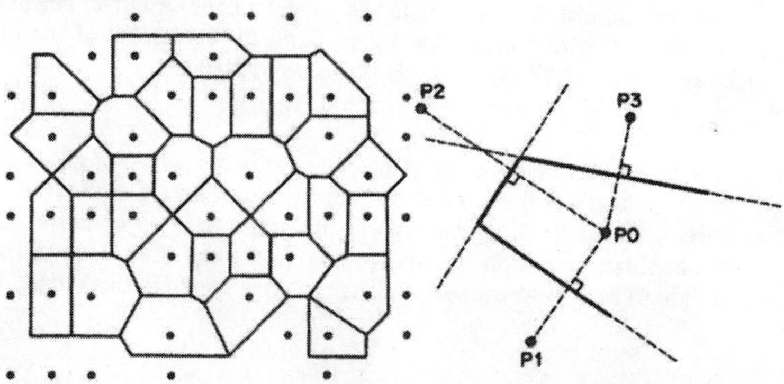

Figure 3.11 A method of defining the area available to each plant when the distribution is irregular. Lines drawn at right angles to lines joining a plant to its neighbours (PO–P1, PO–P2, PO–P3) make the sides of a polygon. The range of sizes and shapes of polygons obtained when there is 50 per cent emergence of squarely arranged seed is shown on the left (drawing I. Currah, NVRS copyright).

further growth of the larger plants is favoured at the expense of the weaker plants. There is sufficient variation in soil in the opportunities afforded for early growth for differences to arise if the plants are genetically uniform.

Uniformity is often desired by horticulturists. In practice, it is more likely to be achieved if we can allocate similar areas to each plant, using a low rectangularity ratio and ensuring that early growth particularly is as uniform as possible.

Population effects

The effects of plant population on yield are best studied by arranging the plants in a pattern which does not intrinsically give different yields according to population. This can be done by using a square, or near square, arrangement. Total dry matter production per unit area was found to be related to plant density as shown in figure 3.2. As plant population increases so does dry matter production until a maximum is reached, at which further increase in population brings about little or no increase in production. The situations in which this relationship does not hold are important in some types of horticulture. In desert farming in arid areas where there is no irrigation, very high plant populations could exhaust the moisture reserves and perish, whereas lower populations would establish deep root systems and persist to give a crop. In most horticultural situations, there are no such limitations.

The type of relationship shown in figure 3.2 can be better defined if a straight line graph can be drawn by expressing the data in some other way. This can be done if we consider the weight per plant as related to number of plants per unit area. If this is plotted we get a concave curve, which can be transformed to a straight line by plotting the reciprocal of the plant weight as in figure 3.12 (a and b). This can be written as

$$w^{-1} = \alpha + \beta\rho \tag{3.3}$$

where w is the mean weight per plant (w^{-1} is the reciprocal of w), α and β are constants and ρ is the number of plants per unit area. α is the intercept and β the slope of the line. We can use this graph to interpolate for yields at any population within the range we used, and (more tentatively) extrapolate to find what the yield might have been outside this range.

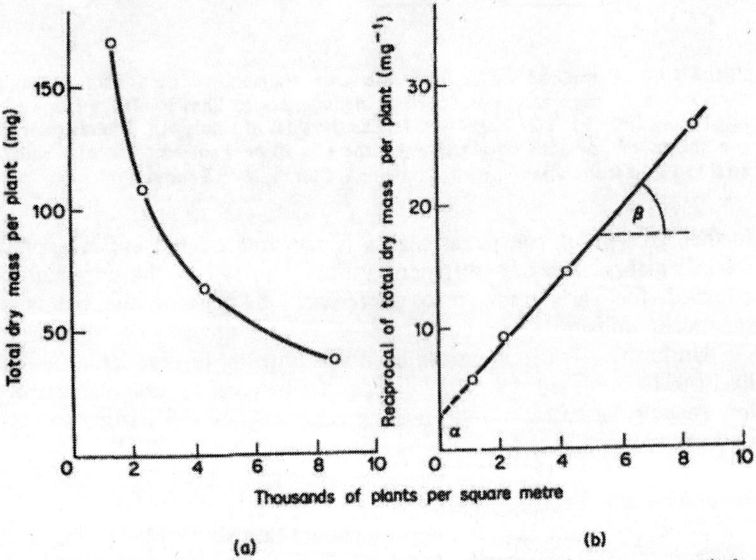

Figure 3.12 The mean dry weight of radish plants plotted against plant population; (a) normal arithmetic scale and (b) a reciprocal scale (α and β are the constants of equation 3.3 given in the text).

Horticulturists are however usually more concerned with the yield of some part of a crop plant: its fruit, bulb or tap root. With some crops the relationship between the yield of this part and plant population is similar to that expressed in equation 3.3. Other crops have some optimum population for the yield of their economically important part. It is very important that horticulturists should know these optimum populations and how they may vary with growth conditions. It seemed that excessive experimentation would be needed before horticulturists could be informed of what populations to use. Systematic designs, similar to the one shown in figure 3.13, were developed and enabled a wide range of spacings to be

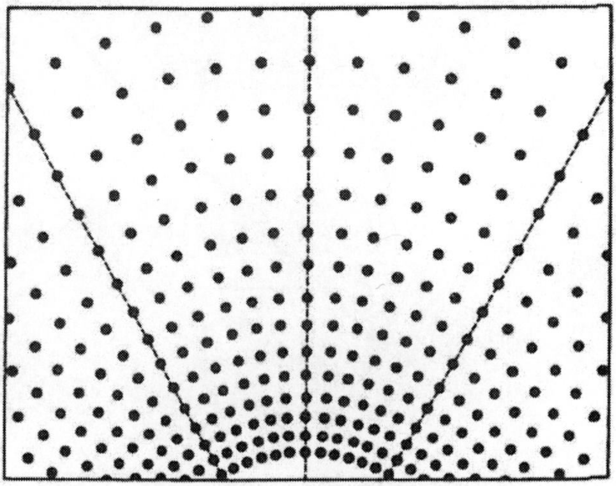

Figure 3.13 A fan design for a spacing experiment with a 'square' plant arrangement. The plant positions are represented by dots.

tested in a small area, furthering our understanding of the relationship between yield and population.

We now realise that competition can affect the proportion of the total production allocated to the economically important plant part. This can again be shown graphically (figure 3.14). When the slope of the line is unity, the proportion allocated to the plant part being considered is not affected by population – so equation 3.3 holds. When the slope is less than unity, the population density is having an effect. If the slope is represented by a symbol θ (which must be less than 1), the equation defining the new situation is

$$w^{-\theta} = \alpha + \beta\rho \tag{3.4}$$

Thus, given the total yield and the yield of the plant part for two population densities, we can draw a graph like figure 3.12, determine θ from the two points, use θ to draw another graph of $w^{-\theta}$ against ρ and get an indication of economic yield at any plant population density.

The accuracy of this approach will depend on the accuracy of the two points and on the validity of the relationships. They appear to be good working relationships but one must always use such mathematical short-cuts with caution.

Competition between crop plants can affect not only total yield, but also the size of the individual plants. Where the relationship between economic yield and population becomes almost constant, there is a wide range of populations where the size of the individual plant is changing but

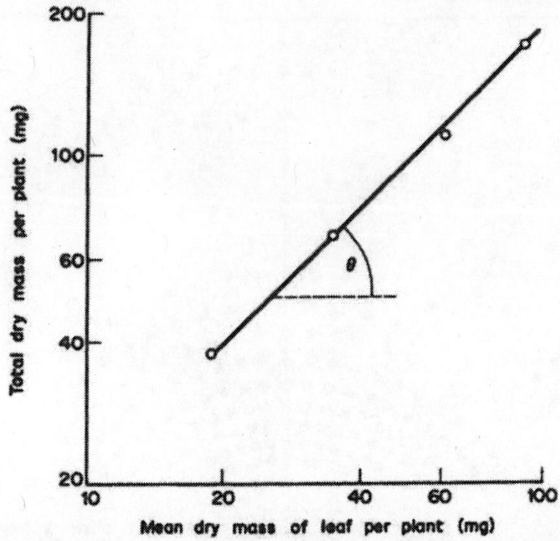

Figure 3.14 When plants differ in size because they have been grown as parts of different populations there is a relationship between total plant weight and the weight of some biological part (for example, leaf or tap root) which is linear when plotted on logarithmic scales as above. Plotted in this way the slope of the line gives an estimate of θ as defined in equation 3.4. (For further discussion see text.)

yield is not affected. Selecting the right population to produce the size required for a specific market is becoming increasingly important. Low populations of carrots give the large vegetable required by food processors making soups or baby food, whereas very high populations give small carrots for canning whole.

Spacing and maturity

Some cultivars of cauliflower will give miniature curds at populations of about 40 per square metre and these have become known as mini-cauliflowers. Competition has almost produced a new vegetable. One characteristic of this closely spaced crop is that nearly all the plants produce their curds at the same time. In conventionally spaced crops of the same cultivars, the curds will mature over a period of several weeks, making selective handcutting the only practicable method of harvest. Mini-cauliflowers with their synchronised maturity, however, can be machine harvested.

We do not understand how competition produces these effects on cauliflower maturity; nor do we understand why high populations of onions ripen earlier than lower ones, or cabbages heart earlier at wider spacing. The effect of plant population on the maturity of such crops as peas is better understood however. At wide spacings pea plants continue to flower and produce pods, so that even when the bottom pods are fully mature there are still flowers on the plant. This situation is ideal for the gardener who finds it convenient to harvest the pods on a cut-and-come-again basis. A high yield is required commercially from one destructive harvest and is achieved by using a high population (90 per square metre) so that each plant has used its resources (space) when only one to two pods per plant have been developed. With this type of crop, continuity of harvesting is generally achieved at wide spacing and single harvesting at very close spacings.

Competition ensures maximum yield, controls plant size and affects maturity. It is one of the major factors affecting crop growth that is within the control of the horticulturist. We now turn to two other such factors.

PRUNING

Pruning is a means of reducing competition between the parts of a single plant, as when branches are thinned out, or between plants, by reducing the size of the competing individuals and so reducing their demand for resources. The optimum population for production of some perennial crops can be maintained by pruning. There is some evidence that in blackcurrants (*Ribes nigrum*), population must be considered not so much as the number of plants per hectare, but rather the number of branches as counted on some arbitrary scale. Young populations planted closely achieve the optimum branch population rapidly, but if the branches are not vigorously thinned in subsequent years, the overcrowded branches give a lower yield of berries. Within broad limits it does not matter if the branches are reduced by removing some plants or by thinning out the branches on each plant.

If overcrowding does not reduce total yield, it will often reduce the size of the plant part being harvested. Pruning can reduce such overcrowding and so overcome this effect. Disbudding of chrysanthemums and camellias reduces the total number of flowers but ensures the formation of larger individual blooms. Pruning that restricts the fruit load on apple and pear trees can similarly increase the fruit size.

Recent work (Jackson *et al.*, 1971) has shown that apples from the inner part of a Cox's Orange Pippin apple tree are smaller, have less red skin and store differently from apples from the outer part of the tree. Shading has similar effects and Jackson *et al.* concluded that differences in storage behaviour made it desirable to pick and store fruit from the inner and outer parts of trees separately. It seems probable that similar differences could be produced by different intensities of pruning.

Great horticultural skill is applied in pruning to control the balance between vegetative and reproductive growth. Such pruning demands an intimate understanding of the normal cycle of development of the plant.

The details of appropriate pruning methods for the many diverse perennial species of horticultural interest cannot appropriately be dealt with here. An example may illustrate the principles, however, and reveal the dangers of pruning without adequate knowledge.

Apples and pears are usually formed from special buds on very short lateral twigs, called spurs. These spurs grow from the lateral buds of a one year old shoot, so that the second season of growth forms a spur with a terminal bud (see figure 3.15). This bud opens the following spring to give leaves and flowers, and is often termed a mixed bud. If pollination follows fruit is formed in the third year. Thus a spur flowers only on alternate years, but on any one tree there are each year some spurs in a flowering phase.

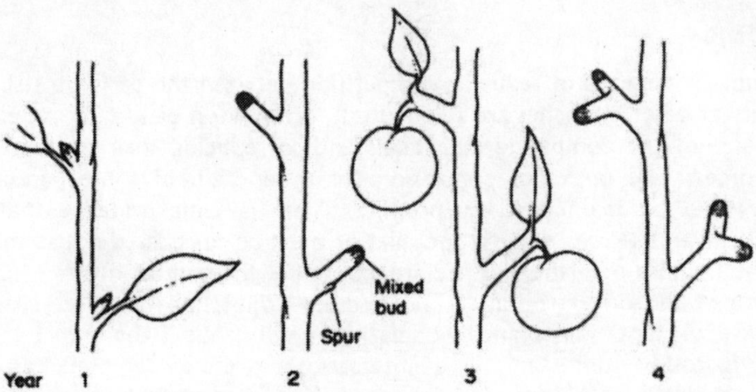

Figure 3.15 The sequence of events leading to spur formation and fruiting in apples. (For further details see text.)

Thus the removal of all one year old wood from an apple or pear tree would effectively prevent spur formation in the next year and reduce fruiting the year after that. The pruning of apple trees and pear trees is normally intended to restrict the number of lateral buds producing spurs by cutting back one year old growths to about six lateral buds, and to thin branches to ensure the strong growth of the wood. Research has established a good positive relationship between branch diameter and ability to produce good size and colour of fruit.

Horticulturists have for long rationalised what experience has shown to produce the required effect. Plant physiologists are slowly learning some of the scientific principles behind horticultural lore and finding that although the observation is correct the rationalisation is less frequently so. For example, it was often observed that newly planted young grape-vines produced shoots that were much less vigorous than those from older plants. This difference in growth was accounted for by assuming that the older plants had greater reserves which the shoots could call upon, but recent experimental work has cast doubt on this.

In Australia, Buttrose and Mullins (1968) grew young grape-vines in water culture and measured shoot growth after imposing four root pruning treatments. These were designed to leave 25, 50, 75 and 100 per cent of the full root system and were obtained by weekly pruning for eight weeks. Within one week of pruning the 25 and 50 per cent treatments already had shoots shorter than the unpruned plants and over the eight week period a clear relationship between root size and shoot size was established. They considered five explanations of this regulation;

(1) supply of water
(2) supply of inorganic nutrients
(3) pruning damage
(4) distribution of photosynthetic products with roots gaining and shoots losing
(5) supply of growth substances from the roots

They advanced reasons for discarding the explanations (1) to (4) and supporting (5), and suggested that cytokinins produced by the roots were what were in short supply. Other work has shown that grape-vine cuttings with roots make good shoot growth, but those without roots will only do so if supplied with cytokinins. This work shows that it could be rash to interpret pruning as a simple shifting of the balance of the supplies of major nutrients.

Apical dominance

It was once contended that the apical bud of a shoot dominates the lateral buds and prevents them from growing because the apical bud is the established sink for a plant's nutrients. In other words, the apical bud develops first and continues to attract nutrient supplies at the expense of lateral buds. This became known as the *nutritive theory* of apical dominance.

In a series of elegant papers starting in the 1920s, Snow produced evidence for a special substance that diffuses from the apex and inhibits lateral buds. These rigorous and skilful experiments left no doubt that the nutritive theory was wrong or, at least, inadequate.

By the mid-1930s other workers had shown that this special substance was indole-3-acetic acid (IAA). If the apex of a shoot were removed and replaced by a source from which IAA could diffuse, then the lateral buds remained just as inactive as if the apex were still present. This simple view of apical dominance did not, however, adequately account for Snow's continued observations. The effects of different concentrations of IAA and the probable role of other hormones are now better understood and this has clarified the correlative inhibition manifest as apical dominance.

Current research is encouraging a return to the nutritive theory however. The use of radioactive tracers has revealed that the hormones making an apex dominant also cause it to be a nutrient sink. Hormones act more rapidly than apical growth in making the apex into a sink.

Recent work by Thomas (1972) indicates that cutting off the apex increases the auxin level in lateral buds and that this is followed by an increase in the level of another plant hormone (gibberellin) as the lateral shoots elongate. Thomas used the ideal material for such a study, namely Brussels Sprouts. (*Brassica oleracea* var. *gemmifera*) with its large lateral buds that make the following of hormone changes practicable.

The horticulturist can see the effects of removing apical dominance every time a hedge is clipped. Frequent clipping reduces apical dominance, encourages lateral buds to grow and gives a close well-set hedge. Less frequent clipping allows fewer shoots to dominate and produces a relatively open hedge.

Apical dominance also plays a vital part in the horticulturist's control over the sprouting of potato tubers. The potato is a swollen stem and the 'rose' end is its apical end. Shortly after harvest the lateral buds (the eyes) are dormant and only the apical bud will grow. Apical growth can be encouraged by keeping the tubers warm (20°C) for 2 to 3 weeks, after which the apical eye will establish dominance over the dormant lateral 'eyes'. The tuber can be stored in this state in cool conditions until near the time for planting when warmer conditions will again encourage the growth of the apical sprout.

Apical sprouting directs all the reserves of the mother tuber into one stem and so improves early yield. If, however, tubers are stored in cool conditions (10°C) from harvest, the apex does not establish dominance before the lateral eyes are released from dormancy. Apical and lateral eyes will then grow to give shoots just before planting, eventually giving a later crop but with a higher yield, because more stems are established in the field for the same planting rate.

Of course, if cut tubers are to be planted, storage that is designed to control apical dominance is not as significant as is the act of cutting in controlling sprout growth.

TRAINING

Pruning is often an integral part of plant training in horticulture and as such it is simply the removal of parts of a plant to achieve a desired shape. This shape usually has no physiological interest or significance – being simply pleasing or convenient. One exception is the horizontal or angled training of young apple trees as espaliers or cordons to induce earlier fruiting.

Wareing and Nasr (1958) grew very young apple seedlings in pots, and some were placed on their sides with the same side of the shoot always towards the ground. New growth was trained horizontally and the pots were only briefly returned to the vertical for watering. These horizontal plants elongated less than normal vertical plants and formed fruiting spurs. Horizontal plants which were repeatedly turned so that different sides faced downwards behaved as though they were vertical. This gravity effect on plant morphology was termed by Wareing and Nasr as *gravimorphism.* It gives some insight into the horticultural training of fruit trees, but the underlying mechanism is not fully resolved. Recent experimental studies (Mullins and Rogers, 1971) support the view that gravity affects the distribution of endogenous growth substances within the apple stems. Lateral buds on the *upper* sides of stems elongated normally if distant from the source of endogenous growth substances — the apical group of buds.

TROPISMS

That plants or plant parts take up or grow into characteristic positions is a commonplace observation: roots grow downwards; shoots upwards; and plants on a window sill bend towards the light. All these plant movements are tropisms in that they are responses to unidirectional stimuli. They are classified according to the stimulus.

Phototropism is the response to light rays. If the plant moves towards the light and aligns itself with the light, the movement is *positive orthophototropism*, as opposed to *negative orthophototropism* which is bending away from light. If the plant organ assumes some other angle to the stimulus it is said to be *plagiophototropic.*

Geotropism is the response resulting from the effect of gravity. The same prefixes are used as in describing the direction of phototropic responses, *positive orthogeotropism* being downward movement on to the line of force.

Haptotropism (also termed *thigmotropism*) is the response to touch or rubbing such as in the coiling of tendrils of climbing plants.

Chemotropism is the response to a chemical gradient, seen in its positive form when pollen tubes grow in a style towards increasing sugar concentrations.

There are other less common tropisms, such as *thermotropism* (response to temperature gradients) and *aerotropism* (response to oxygen). All plant movements are usually classified as tropisms, but tropisms are strictly only those movements related to a directional stimulus. The closing of flowers at dusk and the waving of the tips of climbing plants (*nastic movements*) are not tropisms. The distinction is somewhat academic however, for there is often a close similarity between the mechanisms involved. The lack of directional relationship in nastic movements is more related to the anatomy of the particular plant organ than to its physiology. Tropisms and nastic movements result from unequal growth of cells. Tulip (*Tulipa*) flowers open with rising temperature, which stimulates cell growth on the inside of the petal. Cooling stimulates cells on the undersurface and causes **thermonastic** closure of the flower.

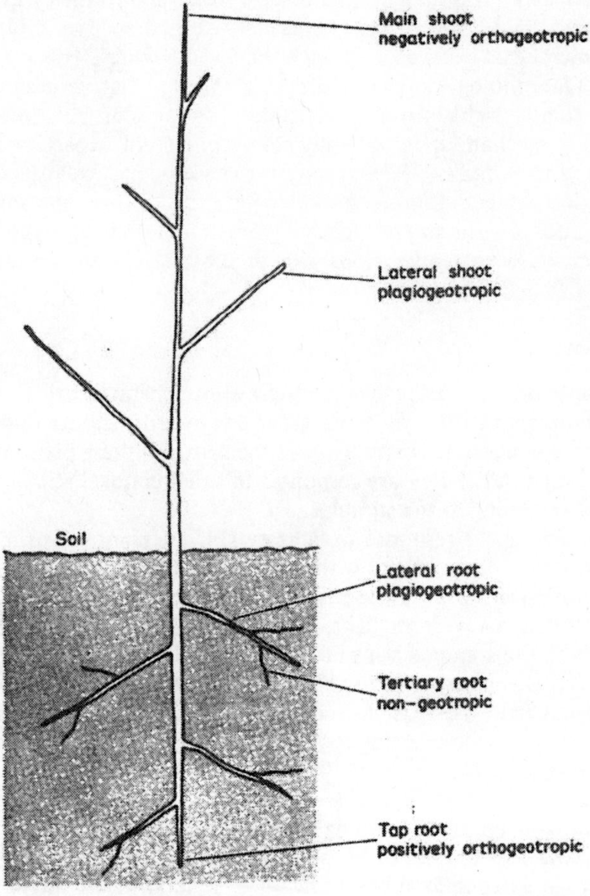

Figure 3.16 Diagrammatic representation of a plant showing how the different tropic responses of organs contribute to the form of the plant.

Tropisms play a major part in determining the form of plants. Consider the diagrammatic vegetative plant represented in figure 3.16:tropisms have influenced its appearance. Organs of a plant can move to assume specific spatial relationships to each other. This **autotropism** is manifest in the movement of leaves to fill gaps in a canopy or in the characteristic angle that leaves maintain to a stem no matter what its position.

Tropisms have been studied for hundreds of years, but a full understanding of the mechanisms involved is still lacking. A current unified theory on the mechanism of phototropism (Curry, 1969) suggests that there is a single photoreceptor, probably a carotenoid pigment, which causes changes in cellular permeability when stimulated. These changes affect, by a poorly understood mechanism, the movement of auxin so that less is found in the cells facing the light. Growth is consequently less on this side than on the other side and so the plant organ curves towards the light. The total amount of hormone from the apex is not affected by light, only its distribution.

Gravity acts in a similar way in that it causes a higher concentration of auxin on the under side than on the top side of both shoots and roots. In view of what has been said about phototropism, this would be expected to induce upward curvature or negative orthogeotropism and this is so in shoots. The positive orthogeotropism of roots must mean that the same relative distribution of auxins as in shoots must cause a downward optimum concentration of auxin for extension. The downward curvature of roots occurs because the concentration is too high on the lower side so there is more extension on the upper side. It has been shown recently that the migration of auxins is paralleled by an even greater movement of gibberellin, which could play a very significant part in causing cell elongation.

Very detailed studies of many aspects of both geotropism and phototropism have been made, and historical accounts of the development, investigation and rejection of theories are fascinating. It was suggested that gravity created differences in the measurable electric potential between the surfaces of a horizontal shoot or root. This in turn led to the observed differences in auxin concentration. Considerable research was needed to demonstrate that it was the differences in auxin concentration that produced the differences in electric potential.

SENESCENCE

As the vegetative plant grows, the oldest leaves begin to senesce. Typically their tips yellow, then their edges and gradually the whole leaf yellows, becomes flaccid and drops off at a time when the next oldest leaves have already begun to senesce. This is termed sequential or *progressive leaf senescence*. When all the leaves die at one time, as with deciduous trees, we have simultaneous leaf senescence. Perennials that die back to the roots each winter exhibit shoot senescence, whereas some annuals which die completely as soon as reproduction is completed exhibit *whole-plant senescence.*

Consideration of these different forms of senescence leads one to realise that leaves, shoots or indeed whole plants are not dying because they are old. Senescence is a controlled process which is often characteristic of the species of plant. This is not to say that either the onset or rate of senescence is unaffected by environment, for shortage of nitrogen in the soil or drought will accelerate senescence. What is important is that we recognise senescence as an inherent process, like growth itself, which is modified by external factors.

The yellowing of leaves indicates a breakdown of chlorophyll, which previously masked the presence of other pigments. Microscopic studies reveal a degeneration of the fine structure, and biochemical tests show that protein content falls as amino acids and amides are exported to other parts of the plant. The breakdown of protein to amino acids and amides occurs in normal green leaves but it is accompanied by their continuous re-synthesis to protein. In a senescing leaf amino acids and amides are exported before re-synthesis can occur. It also appears that other sites are better sinks and so attract the protein building blocks. This theory satisfactorily accounts for sequential leaf senescence (where the younger leaves are sinks) and for **whole-plant senescence** (where the ripening seed has been shown to recover a good deal of the material from the senescing plant).

Concurrent with this loss of protein is the decrease in ribonucleic acid (RNA) content. These losses occur in both detached or attached senescing leaves. Detached leaves accumulate amino acids and amides, suggesting that a senescing leaf is unable to synthesise protein. If a detached leaf is induced to form roots, then both protein and RNA breakdown are slowed down and rooted leaves often survive for a long time, apparently because of the production of cytokinin by the roots. Synthetic cytokinins will delay senescence if applied to leaves, even detached ones. They rapidly increase RNA and protein synthesis and can preserve the fresh appearance of harvested leafy crops for longer than would otherwise be possible. However there are indications that this prolongation only occurs in those cultivars which normally have a short shelf-life. A cultivar with a long shelf-life cannot be improved by applying extra hormones of this type.

In some species, notably dandelions (*Taraxacum officinale*), gibberellins are more effective in delaying leaf senescence than are cytokinins.

Hormones which delay senescence are to be contrasted with others which promote the ageing and abscission of leaves. Abscisic acid is a natural substance that accelerates the yellowing of detached leaves or leaf discs, and causes leaves to absciss. Ethylene produces similar effects and, although not known to be involved in the natural process, ethylene does promote the ripening of fruits (see page 136).

Special layers of cells are usually formed at the base of the petiole to cause senesced leaves to drop off the plant. This thin layer of special cells is at right angles to the axis of the petiole and is called the *abscission layer.* The precise role of auxin, abscisic acid and ethylene in promoting abscission is still uncertain, but although auxin applied to an active leaf will delay the initial stages of its senescence, if senescence is nearing completion then applied auxin will stimulate abscission. Abscisic acid and ethylene promote senescence, making for an earlier abscission, but any other role they may have is more obscure. However, ethephon is used to induce the earlier defoliation of deciduous nursery stock to facilitate lifting for sale.

The synchronous senescence of all a plant's leaves, such as in deciduous woody trees, is stimulated by the environment. The shorter days at the end of the summer, which induce dormancy, coupled with the lower temperatures at this time appear to be responsible for leaf fall. Gibberellins sometimes, but more usually auxins, are effective in preventing this environmentally induced senescence. This has led to the suggestion that the environment is affecting the endogenous auxin levels.

The senescence of leaves is clearly a regulated process, for sequential leaf senescence can prevent the older non-productive leaves lying in the shaded parts of a canopy from becoming 'parasitic'. Synchronous leaf senescence when accompanied by the induction of dormancy enables the plant to pass through an adverse season.

CHILLING AND FROST INJURY

Although growth rate decreases uniformly as the temperature falls within the range 35 to 10°C, the decline may either continue until growth ceases at around 5°C or symptoms of injury may develop (particularly with tropical or sub-tropical plants). At temperatures below 0°C many more plants are lethally injured. On the other hand, the vegetative plants of some species can withstand prolonged periods at very low temperatures. These differences in susceptibility to low temperatures are of considerable commercial and horticultural significance.

The importance of susceptibility to low temperatures in determining species distribution is apparent on macro- and micro-scales. On a macro-scale citrus fruits evidently cannot be grown in temperate regions because they are not cold hardy. On a micro-scale it has been shown, for example, that seedlings of the Lodgepole Pine (*Pinus contorta*) were more tolerant of low temperature in the spring than seedlings of the Ponderosa Pine (*Pinus ponderosa*). In natural stands of Lodgepole Pine in Oregon the minimum night temperature near the soil surface in spring is about −9°C, whereas in Ponderosa stands it is −6°C. This observation, the preponderance of Lodgepole Pine in frost hollows and the sharp transition to Ponderosa Pine on heights above the hollows present strong circumstantial evidence that low temperature susceptibility is a major factor determining the micro-distribution of these two related species. Controlled environment tests showed that all Ponderosa Pine seedlings are killed at −9°C whereas a high proportion survive at −6°C. Lodgepole Pine seedlings do not die until the temperature falls below −9°C.

Chilling injury

Various physiochemical changes have been reported as occurring at low temperatures (above 0°C) in plants of species susceptible to chilling injury. These include

(1) A cessation of protoplasmic streaming

(2) Changes in membranes and enzyme systems that are indicative of a change in metabolic activity
(3) The formation of callose plugs in conductive tissues
(4) A breakdown of chloroplast structure and a consequent yellowing of leaves.

Changes in the pattern of metabolic activity have also been reported to occur at low temperatures in plants that do not show chilling injury, such as *Pisum sativum*. In susceptible species, however, these changes are demonstrably harmful and sometimes irreversible.

Plants can be 'hardened' to withstand temperatures that would otherwise cause chilling injury. Imbibed cucumber seeds that are exposed for 12 hours each day to temperatures of between 0 and 2°C for 10 to 15 days before germination give more hardier plants. The chloroplasts of cucumber plants, previously hardened by growth at temperatures near 12°C, were not disrupted after exposure for several days to temperatures between 3 and 12°C, whereas the chloroplasts of unhardened plants were gradually disintegrated by this cold treatment.

In addition to hardening plants by exposure to low but non-injurious temperatures, similar or even more marked hardiness can be achieved by growing the plants in an atmosphere containing less oxygen than normal. Cucumber seedlings grown from seed in an atmosphere of only 2 per cent oxygen survived freezing at −10°C, whereas air grown seedlings all perished. In general, anything that reduces growth is liable to produce a hardening effect.

Chilling injury that produces no visible symptoms can nevertheless adversely affect growth. Chilling of germinating cotton seed reduces plant height, delays fruiting and reduces fibre quality. Chilling injury resulting in the yellowing of leaves of *Phaseolus vulgaris* can prove lethal, but some cultivars show some recovery when the temperature rises again, and almost normal growth can be ultimately resumed. The check to growth is likely to delay maturity and reduce yield. Some other species of *Phaseolus* are more resistant to chilling injury than *P. vulgaris* and may prove useful sources of hardiness for breeders.

The study of chilling injury and stress is a relatively recent challenge for plant physiologists, and further elucidation of the mechanism involved should prove rewarding.

Frost injury

At temperatures slightly below 0°C, ice begins to form inside the plant. Various factors affect the precise temperature at which this happens and some of these will be mentioned later. Ice formation is not necessarily harmful because it occurs in both hardy and non-hardy species. Furthermore the extent of the damage done depends not only on the species of plant, but also on such factors as the age of tissue, and the degree of acclimatisation to low temperature before freezing.

Just over one hundred years ago Sachs wrote about ice formation in plants.

> When ice is formed in the tissues of a plant, two points must be taken into consideration. The water, when about to freeze, is on the one hand contained in a mixed solution, the cell-sap; on the other hand it is retained by the force of cohesion as water of imbibition in the molecular pores of the cell-wall and of the protoplasmic bodies. Now it is an established fact of physics that a solution when freezing separates into pure water which solidifies into ice and a concentrated solution with a lower freezing point. A portion of the cell-sap-water becomes therefore by freezing more concentrated . . . . How far this circumstance must be considered in the destruction of cells by freezing and thawing is yet to be decided.

He went on to say that all this can be observed by freezing a starch-paste, which has 'the appearance after thawing of a spongy coarsely porous structure, the water running off clear from its large cavities'. Sachs' observations are still valid today, as indeed are many of his other statements on this subject.

The freezing of cells

The dead cells of xylem vessels contain the most dilute solutions of solutes within the plant. As a consequence this solution shows the minimum depression of freezing point and, as the temperature falls, ice is first formed within these vessels. Their walls form a continuum with the walls of live cells. Water within this ramifying system of cell walls is the next to freeze, having been 'seeded' by the ice formation within the xylem.

Each living cell of a plant behaves as though there is, enclosing the cytoplasm and other cell inclusions, a membrane which normally lies tightly against the inside of the cell wall. When the water in the cell wall freezes, the vapour pressure is reduced and water moves through the cell membrane owing to this new gradient and more ice is deposited on the cell wall. The cell contents, still enclosed by the membrane, shrink away from the walls and are concentrated, thereby lowering their freezing point. This sequence of events is illustrated in figure 3.17, which also shows that in non-hardy cells ice forms within the cytoplasm. This ice formation causes mechanical damage to the essential structures of the cell and eventually causes cell death.

The rate of freezing and thawing

If the temperature falls very rapidly, water cannot migrate quickly enough through the cell membrane to depress the freezing point of the cytoplasm sufficiently to prevent ice forming within it. In these circumstances the hardy cell behaves essentially like the non-hardy one.

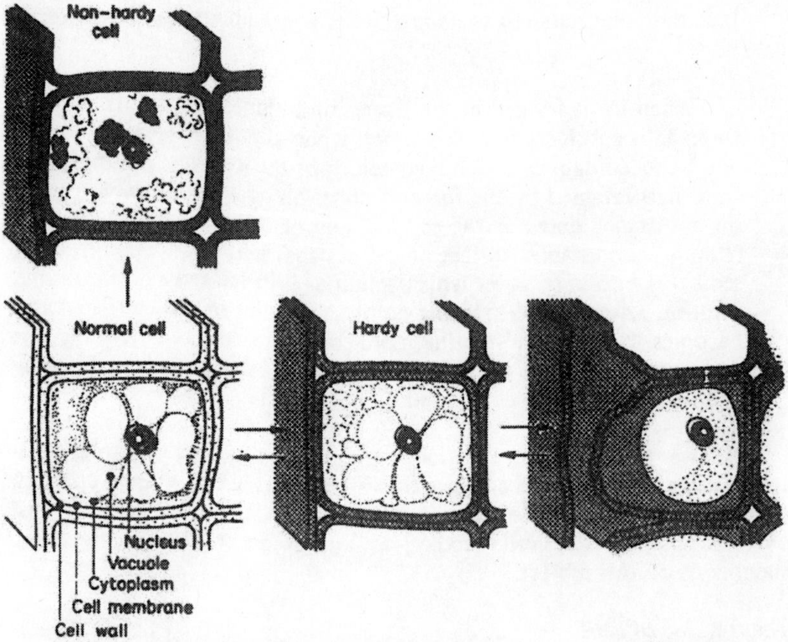

Figure 3.17 Diagrammatic representation of the progressive freezing of plant cells. When ice forms around a cell the cytoplasm gives up water and solutes to it. Although inside the cell wall, the ice (stippled area) never penetrates the cell membrane. This is the sequence in hardy cells and is termed equilibrium freezing. In non-hardy cells ice breaks through the membrane and disrupts the cell contents (from Idle, 1968).

Thawing of a non-hardy cell simply results in tissues composed of disorganised and dead cells. The plant collapses and the tissues rapidly decompose. If hardy cells are thawed slowly, then water from melting ice is taken back in through the membrane and the cytoplasm returns to its original state. If thawing is rapid – as often happens when the sun shines in a clear sky following a frosty night – the ice melts faster than the membrane can cope. The plant becomes 'flooded' and water drains away from some cells and even may escape from the plant through cracks in the epidermis. This redistribution or loss of water can prove lethal.

'Glacier formation'

If the cytoplasm within the cell membrane of hardy plants contracts too far away from the cell walls, then there is a danger that the plasmodesmata (the strands of cytoplasm that interconnect cells) will be snapped. This is avoided by the cell walls also contracting (as shown in figure 3.17) to give large cavities between cells where ice can form. Ice formation cannot occur between all cells because this would also disrupt the plasmodesmata. To be effective, a large cavity serving blocks of tissue must form. These large cavities were termed 'glaçons' by Prilleaux in 1869, and the entire process is known as *glacier formation.*

Because ice is first formed in the vascular tissue, it would be reasonable to expect glaçons to form within the vascular system. In non-hardy plants the vascular system is indeed disrupted by freezing, but the glaçons form in hardy plants at sites in the tissue where little or no disruption is caused. It appears that in hardy plants these sites are characterised by the presence of a few susceptible (easily frozen) cells which form a nucleus for glacier formation. One of the characteristics of hardy plants, therefore, is the possession of small groups of 'non-hardy' cells at sites where mechanical damage is not too detrimental. In non-hardy plants glacial formation occurs more indiscriminately and damages vital tissues.

Hardening and hardiness

At a cellular level the main difference between hardy and non-hardy plants lies in the permeability of their cytoplasmic membranes. When hardy species are subjected to low temperatures (but above 0°C), their cytoplasmic membranes become more permeable as they become more able to withstand freezing. This permeability effect is not seen in non-hardy plants, nor do sugars and proteins accumulate in their cells as in hardy species.

Many other biochemical changes occur during the hardening of plants, but their significance is controversial. For example, Levitt (1962) has suggested that when a cell freezes and water is separated from the cell contents, injury is mainly caused by dehydration. He postulated that plants differ in their ability to withstand freezing largely because of differences in their ability to withstand dehydration. As dehydration proceeds the sulphydryl (SH) and disulphide (SS) groups on adjacent protein molecules approach each other more closely and chemically combine. These bonds are strong and do not break when water re-enters the cell during thawing, as do weaker bonds within the proteins. The proteins consequently become denatured, causing irreversible damage. It has been suggested that this damage is prevented in frost-hardy species because sulphydryl groups are protected by substances which prevent these strong bonds forming.

Protection against freezing injury

I have already referred to the horticultural practice of 'hardening' plants by exposure to low temperature (but above 0°C) to make them more able to withstand freezing. Apart from the physical protection afforded by the use of glasshouses or other shelters, 'hardening' is probably the most generally adopted method for ensuring survival of crops during cold weather. We have also seen that rapid thawing can cause more damage than freezing, and this explains why horticulturists do not plant susceptible species where the early morning sun falls.

Damage from freezing can be greatly reduced by the formation of ice on the outside of the plant. Fortunately, this occurs normally because dew formation precedes night frosts and the dew freezes before the xylem sap. If the surface of the plant is dry, the sap is likely to super-cool and consequent ice formation is sudden. Sudden formation of ice occurs within the cells of both hardy and non-hardy plants and results in cell death. Ice on the outside of the plant acts as a nucleus for internal ice formation and prevents this lethal super-cooling.

Sprinkling plants with water as an anti-frost measure undoubtedly owes some of its success to the prevention of super-cooling, but also of importance is the heat given up as water turns to ice. Many experiments have probed the mechanics of protecting plants in this way and have shown the importance of precipitation rates and droplet size. Rogers *et al.* (1954) found that, to protect apple blossom, sprinkling may be delayed until shortly after the air has reached 0°C and, provided the wind speed was 2 m.p.h. or less, 2 mm of water per hour was likely to protect the buds down to −3.3°C, 4 mm down to −4.8°C and 6 mm down to −5.8°C.

Several chemical sprays have been tested for possible frost-protection properties. Synthetic growth inhibitors, auxins, gibberellins and cytokinins have all been shown to play some part in enhancing frost resistance. Whether they act directly is a matter of considerable doubt, but it seems likely that the hormones must be properly balanced. The dormant state of buds is under hormonal control (page 99) and confers considerable frost-hardiness. Why exactly this should be so is not known, but it is thought to be because of changes in the characteristics of the protoplasm and in the permeability of the cell membranes.

The use of growth inhibitors to delay bud break, and so avoid damage by spring frosts, has also been investigated. The side effects of the inhibitors, however, are usually so detrimental that on balance it is better to risk the frost damage.

References

BROWN, D. M. (1969). Heat units for corn in Southern Ontario. Ontario Department of Agriculture and Food Agdex, No. 111/31.

BUTTROSE, M. S., and MULLINS, M. G. (1968). Proportional reduction in shoot growth of grapevines with root systems maintained in constant relative volumes by repeated pruning. *Aust. J. biol. Sci.*, 21, 1095–101.

CURRY, G. M. (1969). Phototropism in *Physiology of Plant Growth and Development* (Ed. M. B. Wilkins). McGraw-Hill, London. pp. 245–73.

DONALD, C. M. (1961). Competition for light in crops and pastures. *Mechanisms in Biological Competition* (Ed. F. L. Milthorpe). University Press, Cambridge, pp. 282–313.

GRAY, D. (1981). Predicting the dates of harvest and reliability of cropping of outdoor bush tomatoes in the U.K. *Acta Hortic.*, No. 122, 133–40.

IDLE, D. B. (1968). Ice in plants. *Science J.*, 4 (1), 59–63.

IKUSIMA, I., and KIRA, T. (1958). Effect of light intensity and concentration of culture solution on the frond multiplication of *Lemna minor* L. *Seiro-Seitai*, 8 (1), 50–60.

JACKSON, J. E., SHARPLES, R. O., and PALMER, J. W. (1971). The influence of shade and within-tree position on fruit size, colour and storage quality. *J. hort. Sci.*, 46, 277–87.

LEVITT, J. (1962). A sulphydryl-disulphide hypothesis of frost injury and resistance in plants. *J. theoret. Biol.*, 3, 355–91.

METEOROLOGICAL OFFICE (1971). Tables for the evaluation of daily values of accumulated temperature above and below 6°C from daily values of maximum and minimum temperature. (Revised and to be read in conjunction with Form 3300, 1959.) Meteorological Office, Bracknell.

MULLINS, M. G., and ROGERS, W. S. (1971). Growth in horizontal apple shoots. Effects of stem orientation and bud position. *J. hort. Sci.*, 46, 313–21.

NELDER, J. A., AUSTIN, R. B., BLEASDALE, J. K. A., and SALTER, P. J. (1960). An approach to the study of yearly and other variations in crop yield. *J. hort. Sci.*, 35, 73–82.

NIETO, J. H., BRENDS, M. A., and GONZALEZ, J. T. (1968). Critical periods of the crop growth cycle for competition from weeds. *Pestic. Abstr. C.*, 14, 159–66.

ROGERS, W. S., MODLIBOWSKA, I., RUXTON, J. P., and SLATER, C. H. W. (1954). Low temperature injury to fruit blossom. IV Further experiments on water-sprinkling as an anti-frost measure. *J. hort. Sci.*, 29, 129–41.

SACHS, J. (1875). *Textbook of Botany; Morphological and Physiological*. Clarendon Press, Oxford.

SNOW, R. (1940). A hormone for correlative inhibition. *New Phytol.*, 39, 177–84.

THOMAS, T. H. (1972). The distribution of hormones in relation to apical dominance in Brussels sprouts (*Brassica oleracea* var. *gemmifera* L.) plants. *J. exp. Bot.*, 23, 294–301.

WAREING, P. F., and NASR, T. A. A. (1958). Gravimorphism in trees. Effects of gravity on growth, apical dominance and flowering in fruit trees. *Nature, Lond.*, 182, 379–81.

WILLIAMS, R. F. (1946). The physiology of plant growth with special reference to the concept of net assimilation rate. *Ann. Bot.*, 10, 41–72.

Further Reading

BLEASDALE, J. K. A. (1966). Plant growth and crop yield. The fourth Barnes Memorial Lecture. *Ann. appl. Biol.*, 57, 173–82.

MAYLAND, H. F., and CARY, J. W. (1970). Frost and chilling injury to growing plants. *Adv. Agron.*, 22, 203–34.

WAREING, P. F., and COOPER, J. P. (Editors) (1971). *Potential Crop Production*. Heinemann, London.

CHAPTER 4

VEGETATIVE PROPAGATION

The vegetative plant of most species can reproduce without flowering and forming seed. Reproduction of this type is termed vegetative propagation and is often the main way a plant multiplies. Vegetative propagation is of great horticultural significance as it enables the faithful reproduction of plants that either do not breed true from seed or are sterile. Vegetative propagation always produces plants genetically identical to the parent. A group of plants that is vegetatively derived from one parent is termed a **clone**, no matter how many generations of vegetative reproduction were involved in producing the group.

NATURAL VEGETATIVE PROPAGATION

Natural layering

When some plant stems are lying in a horizontal position for some time, perhaps as a result of being blown over, roots begin to form on the under surface. This will occur even if only part of the stem is horizontal; roots will form under the horizontal part of a kink in a stem. If a horizontal tomato plant is slowly but continuously rotated on a klinostat, roots do not form; but if one side remains as the lower side, then roots form within a few weeks. Apparently, gravity has induced a substance to migrate, so that it is in higher concentration on the lower side of horizontal stems.

This is comparable with the phenomena of geotropism (page 77) in which the migrating substance is auxin. It is now well established that auxins stimulate root formation and can be obtained as proprietary rooting powders. We will return to this aspect of the physiology of propagation later in this chapter.

The roots induced on stems do not arise from other roots, and so are termed **adventitious.** They arise from undifferentiated but actively dividing cells deep within the stem. To emerge they must rupture overlying tissues. In some species (notably the carnation) the tissues are not easily ruptured so the roots turn downwards and grow to emerge from the bottom of a cutting.

Wounding the outer layer of a stem will often weaken the constraint of these outer tissues and produce reactions within the plant which encourage root growth (see figure 4.1). Sufficient wounding can be caused simply by pushing a cutting into the sharp sand, which is often favoured

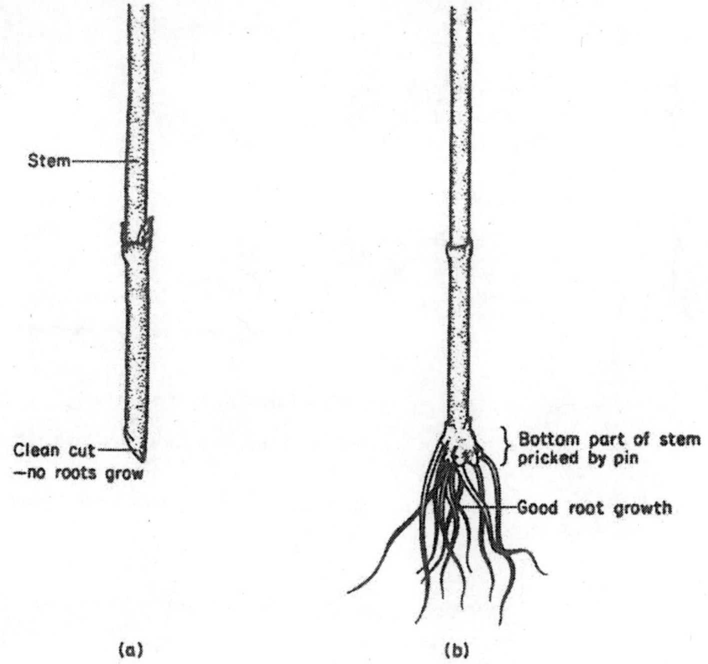

Figure 4.1 Privet (*Ligustrum vulgare*) cutting after 3 weeks in water; left, base of cutting taken with a clean cut; right, similar cutting except that the basal centimetre was scratched with a pin.

as a rooting medium. Wounding and the induction of rooting by gravity can play a significant part in natural layering of plant parts that are not particularly adapted to furthering the vegetative propagation of a species. With many species, however, modifications of the plants lead to particularly effective natural layering which is exploited by horticulturists.

Stolons

The 'runners' produced by strawberries and *Saxifraga sarmentosa* are specialised stems. Examination of a strawberry runner (figure 4.2) reveals that one new plant arises from every other node on the elongated branch that emerges from the mother plant. This branch is called a stolon, and plants formed on it may in turn produce their own runners.

In most strawberry cultivars the formation of runners is under **photoperiodic** control — the length of the light period during a 24 hour day governs runner formation. Stolons are produced only when the day length exceeds 12 to 14 hours, depending on the cultivar. As we shall see in chapter 5, the duration but not the intensity of light is important here, quite low intensities from ordinary incandescent bulbs being all that is

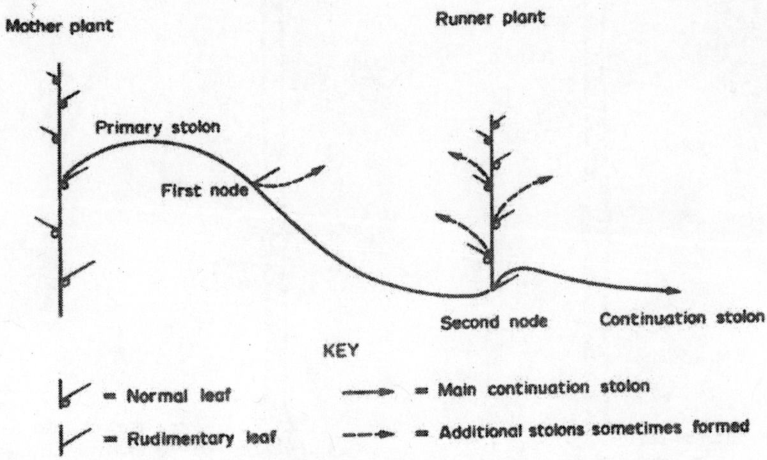

Figure 4.2 Diagram showing the development of a primary stolon and runner plant of strawberry (from Robertson and Wood, 1954).

needed. De-blossoming plants that are intended for runner production increases by about 30 per cent the number of stolons growing directly from the mother plant (Robertson and Wood, 1954).

In certain species stolons may arise underground, their tips either eventually turning upwards to give a shoot, or becoming modified to form a specialised perennating organ (as in the potato). As with organs above ground, the formation of these underground organs in some species is under photoperiodic control.

Rhizomes

Rhizomes are also specialised underground stems which are important for the vegetative propagation of some plant species, including (unfortunately) many troublesome weeds.

Physiologists have studied the vegetative propagation of *Agropyron repens* (Couch Grass) in order to control its growth. These studies have shown that fragmentation of the rhizomes by cultivation breaks the dormancy of the rhizome's buds. Usually only one bud on each rhizome fragment will develop into a shoot, for its growth suppresses all other buds. These buds can remain dormant until some further cultivation kills the dominant shoot. When new shoots have two leaves and have not yet tillered, about 70 per cent of the carbohydrate reserves have been utilised and are at their lowest ebb. Further cultivation for weed killing will then have its maximum effect, but any delay allows new rhizomes to form quickly and reserves are built up. In the low light intensities of temperate winters, some of the carbohydrate reserve is lost. Thus, cultivation aimed at producing two-leaved shoots just as winter starts will considerably weaken the weed. If the rhizome is then buried at least 15 cm deep, the reserves are too low to allow re-emergence. More effective ways of killing couch grass using herbicides are often combined with autumn cultivation for maximum effect.

Suckers — shoots arising from underground

The roots of many species of plants have properties similar to rhizomes in so far as they can at some distance from the mother plant produce shoots, which in due course become independent plants. These shoots arise from spontaneous buds on the roots and, because these buds do not derive from stem tissue, they are called adventitious buds.

In some species the normal shoots have a limited life and all new shoots arise from spreading roots. Familiar examples are *Kerria japonica* and Red Raspberry (cultivars of *Rubus idaeus*). Studies made to improve the horticulturist's ability to multiply the raspberry from root cuttings indicated that the roots seem unable to produce new shoots during the summer months. This seasonal dependence of suckering and the ability of root or stem cuttings to produce plants is well known in horticulture.

Offsets

Many bulbs and corms reproduce by what horticulturists call offsets, which are small bulbs (or corms) clustered at the base of the mother bulb. The term offset is also used to describe shoots, usually rosette-like, which arise around the mainstem. As the term implies they usually represent an easy means of plant multiplication for the horticulturist. Again their formation is often seasonal, suggesting that seasonal changes are somehow influencing the parent plant to produce them. This is clearly so in the case of Chrysanthemum (see page 118).

Physiological treatments can sometimes induce offset formation, even where none may occur naturally. For example, the Bulb Onion (*Allium cepa*) does not normally reproduce vegetatively. If however the flower buds are cut off the seeding head to leave a 'ball of bristles', then small bulbs form in between the 'bristles'. The number of bulbs formed can be greatly increased if the 'shaved' flower head is sprayed with hormone (see figure 4.3).

In Daffodil (*Narcissus* sp.) bulbs, a new bulb grows within the old one each year. After flowering more than one growing point may become active and two or more bulbs arise within the old scales. One of these bulbs is usually bigger than the others but throughout two seasons the smaller ones become large enough to be broken away as independent bulbs. Tulips (*Tulipa* sp.) form new bulbs each season, arising from axillary buds between the scales of the parent bulb which disintegrates after flowering.

The axillary buds between the scales of hyacinth (*Hyacinthus* sp.)

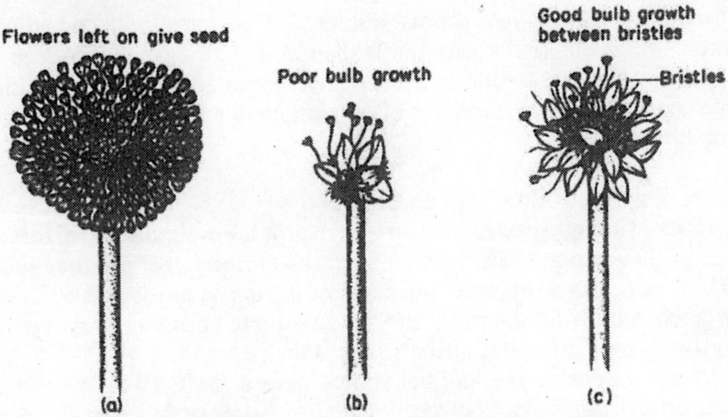

Figure 4.3 Stimulation of bulblet development on clipped inflorescences of onion cv. Rocket by 6,N-benzyladenine (BA). (a) Normal inflorescence; (b) clipped inflorescence sprayed with water; (c) clipped inflorescence sprayed with BA solution (T. H. Thomas, NVRS copyright).

bulbs can be stimulated to produce small bulbs by removing or damaging the dominant apex of the parent bulb. Scooping out the main growing point is done horticulturally to promote bulblet formation to multiply new cultivars more rapidly than would otherwise be possible. Damage by insects or other natural causes can produce similar effects; the mechanism ensures survival if the parent bulb cannot grow and flower.

Bryophyllum daigremontianum forms small plants in the notches along the serrated margin of its leaves when grown in long periods of light. These plantlets drop from the parent and quickly establish themselves as new independent plants. Plants of this species are among the most rapid at forming roots on the undersurface of horizontal stems.

PLANNED PROPAGATION

The reader will have realised that horticulturists plan propagation often by following or enhancing some natural form of multiplication. This is not true however of the most usual form of vegetative multiplication used by gardeners, that is propagation from cuttings.

Cuttings

Pieces of stem lacking leaves or buds do not root readily and often not at all. This observation was made nearly one hundred years ago and led to the conclusion that active buds or young leaves above the cut somehow supply something that stimulates root development. Went and Thimann (1937) in 1935 showed that indole-3-acetic acid (IAA) would stimulate rooting and concluded that this natural hormone (or some other similar hormone) was the key substance.

Synthetic auxins such as naphthaleneacetic acid (NAA) and indole-3-butyric acid (IBA) produce similar effects. Some species seem however to need some substances in addition to natural or synthetic auxin to form roots. The general term *rhizocalines* is the name given to these unknown compounds, and examples of plants that seem to need rhizocalines are many coniferous trees, *Hibiscus* sp. and the Mung Bean (*Phaseolus aureus*). Working with the Mung Bean, Hess (1969) showed that neither IAA nor NAA significantly increases root formation on cuttings. Catechol, typical of similar organic compounds occurring naturally in plants, stimulates root formation more than auxins, and when applied together with IAA or NAA the combined effect was greater than expected. This type of bonus action is termed **synergism**. Hess suggested that catechol protects the IAA and NAA from destruction by enzymes in the cuttings. Thus his theory does not involve any other substance actually stimulating root initiation.

Root fragments of some species will form adventitious buds to regenerate a whole plant, and there is some evidence that the formation of these buds depends on the presence of appropriate concentrations of cytokinins.

The leaves of some species, notably *Begonia*, will form both adventitious roots and buds, numerous plants being produced from a single leaf if it is laid on moist sand and its veins are cut across. Wounding stimulates cell division, possibly because specific wound hormones are produced (although there is only one instance of such a hormone having been isolated). The actively dividing cells that arise as a result of wounding form the callus (or proliferation of undifferentiated cells) often seen at the base of a stem cutting. Either within this callus or at sites within the parent tissue some of these dividing cells become organised as the initials of new buds or roots. Hormones play a vital role in determining the nature of this differentiation.

Polarity

Cuttings exhibit polarity in that roots appear at what was formerly the lower end. Root cuttings of Chicory (*Cichorium*) form shoots at the end that was formerly nearest the stem (proximal) and roots at the other end (distal). This will occur even if the root cuttings are placed horizontally or upside down in the rooting medium. Segments of twigs of willow (*Salix* sp.) readily grow roots if held in moist air, and so are ideal for demonstrating polarity, for even if the twig is suspended upside down in a moist chamber, roots form at what was formerly the lower end.

Polarity is an inherent property but poorly understood. It has been suggested that the micro-structures of the cell walls are aligned in a manner analogous to the alignment of the atoms in a bar magnet, but there is no evidence to support this. Auxin moves downwards in a stem cutting and it is reasonable to suppose that this may be the basis of polarity. If auxin is applied to the upper end of a cutting, rooting is stimulated at the lower end.

The nutrition of the parent plant from which a cutting is taken can effect the likelihood of subsequent root formation. Krauss and Kraybill (1918) were among the first to observe this phenomenon. They showed that cuttings taken from yellowish stems of tomato plants had much carbohydrate but little nitrogen and as cuttings produced many roots, but gave poor shoot growth. On the other hand, cuttings from greenish stems that were high in both carbohydrates and nitrogen produced fewer roots but had strong shoots. Very succulent green stems, low in carbohydrate but high in nitrogen, rotted before they rooted. Working with cuttings of pine (*Pinus sylvestris*) seedlings Hansen *et al*. (1978) showed that higher light intensity for the stock plants was more important than higher light for the cuttings in ensuring rooting. They attributed the difference to the higher carbohydrate content of the cuttings.

Iodine, which gives a dark blue coloration in the presence of starch (a common carbohydrate), can be used to select good cuttings. If the cut ends are put into a 0.2 per cent solution of iodine in potassium iodide for 1 minute, those giving the darkest colour are to be favoured. An experiment with vine cuttings gave 63 per cent take for the high starch cuttings, 35 per cent for the medium and only 17 per cent for the low (Winkler, 1927).

The age of the parent plant

Plants that are difficult to root may not necessarily have the wrong carbohydrate/nitrogen balance. Other factors are involved and can be of over-riding importance. For example, cuttings from seedlings usually root more readily than those from older plants. This *juvenility factor*, although its physiology is not understood, is frequently important for species that are difficult to root, such as some species of conifers.

Age probably has its effect through its influence on the nutritive content of the cutting. Flowering and non-flowering shoots of difficult plant species also differ in their rooting ability, probably for the same reason. Non-flowering material roots more readily.

Time of year

The season of the year at which cuttings are taken affects their success. The reasons for this may be more related to the weather that the cutting subsequently encounters rather than to any inherent property of the cutting. Further, from what has been said about the effects of the nutritional status of cuttings, it follows that at certain seasons it will be easier to find the right balance of nutrients. Auxins are produced by active buds and young leaves so cuttings without leaves and with dormant buds will usually fail to root. However when all considerations of this kind have been taken into account there remain some species in which seasonality is unaccounted for; for example, Chinese Fringe Tree cuttings (*Chionanthus retusus*) are best taken in mid-spring, whereas narrow-leaved evergreens such as *Juniper* sp. root better if cuttings are taken during the winter.

The root-promoting properties of auxin and synthetic auxins only help to overcome some of the inherent difficulties encountered in trying to root cuttings. Rooting can be inhibited if auxins are applied at the wrong concentration. If a proprietary compound is being used the only guide to the right concentration is that given by the manufacturers. It is folly to try to improve performance by using more than is recommended. Furthermore, many hormone preparations are unstable and it is usually false economy to try to use the same pack of rooting powder for more than one season. Strong sunlight can destroy a 10 p.p.m. solution of IAA in 15 minutes, but IBA is much more stable. Fresh preparations should be used whenever possible and the extensive guides to suitable methods of treatment should be consulted (see references at end of chapter).

DORMANCY

Bulbs, corms and tubers are familiar forms of vegetative propagation. They are organs that are adapted to ensure survival through an adverse period and provide the reserves for renewed growth, while at the same time increasing the number of individual plants produced.

Because of their inherent ability to withstand adverse conditions many bulbs, corms and tubers are easily stored and have consequently become important food crops (for instance, potatoes and onions). Others have become horticulturally important because of their flowers. The corms of *Crocus* sp. and the bulbs of *Narcissus* sp. are stored and traded in vast quantities each year.

In these vegetative storage organs growth is temporarily suspended: they are dormant. To understand how best to treat these organs to achieve some desired effect we must first consider what is known about dormancy.

Seed dormancy was described in chapter 1 along with some techniques for overcoming dormancy. We saw how seed can exhibit different kinds of dormancy at different seasons of the year (page 5). With vegetative storage organs, dormancy is usually an inherent property not induced by adverse conditions. It is therefore termed **innate dormancy**.

It seems reasonable to suppose that the normal metabolic processes are either inoperative or reduced to a very low level during innate dormancy. Clues as to how this occurs have come from studies on the effects of environmental factors which induce, prevent or break dormancy.

The induction of dormancy is commonly under photoperiodic control. Deciduous trees such as birch (*Betula pubescens*) and larch (*Larix decidua*) will not form dormant resting buds if held in long days in warm conditions. Although other factors such as falling temperature are involved, they do not have such a general or dominant effect as photoperiod. This is probably of evolutionary significance.

Adverse seasons in any climatic region will be most consistently heralded by a change in day length. Temperature changes or variations in water availability are nothing like as consistent. A plant that responds to changes in day length would consistently be ready for the season to come. Dormant buds on above ground shoots are typically adapted to withstand the low temperatures and water stress (induced by the lack of availability from frozen soils) of a temperate winter. The most consistent prelude to winter in temperate regions is the reduction of day length, so we find bud dormancy in temperate tree species associated with short days. Underground storage organs are protected during a temperate winter, and dormancy is again induced by short-days in, for example, rhubarb root stocks. In other species the underground storage organ appears to have been adapted to survive a dry period that is heralded by long days. Thus we find that onion (*Allium cepa*) becomes dormant in long days.

Dormancy is probably caused by the presence of an inhibitor rather than by the lack of some promoter. There is increasing evidence that one of the principal inhibitors is abscisic acid (ABA), which may act as it does in the promotion of senescence (page 80) by inhibiting RNA and protein synthesis. Dormancy is most commonly removed by a period of low temperature. For example, dormant rhizomes of Lily-of-the-Valley (*Convallaria majalis*) require about two weeks at 2°C to remove their dormancy. There is evidence from several species to support the view that chilling or other treatments which break dormancy also reduce the levels of inhibitor. At the same time as, or closely following, inhibitor loss the levels of gibberellins increase, but this seems to be the result rather than the cause of dormancy removal. The evidence for this last statement is scant however. In experiments with dormant rhubarb root stocks I found that massive doses of gibberellin induce growth for only short periods. This suggested that when the gibberellins were used up the dormant state was still fully present. It was as though dormancy had not been broken but that I had provided a limited supply of something that the plant produces itself once dormancy is broken. Treatments with ethylene chlorhydrin (see page 100), on the other hand, did break dormancy permanently: presumably the natural mechanism had been triggered.

Subsequent work by others has shown that gibberellic acid can usefully promote the growth of rhubarb provided part of the normal dormancy breaking by low temperatures has been completed. One of my colleagues at Wellesbourne, Dr Tudor Thomas (1969), has shown that gibberellic acid will not induce dormant onion bulbs to sprout, but once sprouting has started it appears to promote shoot development.

The length of the period at low temperature required to break dormancy depends on many factors. *Gladiolus* corms only need 24 hours at 0 to 5°C to break their dormancy, whereas some cultivars of onion may need several weeks at 0 to 5°C before they are ready to sprout. There is ample evidence that cultivars can differ in their response to cold and there is some evidence to suggest that the conditions under which dormancy is induced can also affect the cold requirement. In temperate conditions, however, it is not uncommon for this cold requirement to have been fully satisfied well before mid-winter. This enables us to lift rhubarb roots for forcing or to cut twigs of *Forsythia* that will open in a vase in the house.

THE STORAGE OF PERENNATING ORGANS

That conditions of storage can affect the subsequent growth of bulbs, corms, tubers and rhizomes will be evident from what has been said about dormancy. Low temperature is the most significant environmental factor in overcoming dormancy.

It follows that storage at relatively high temperatures should lead to a prolongation of dormancy. At Wellesbourne we experimented to see if a practical system of onion storage at high temperatures (about 25°C) could be devised. Although the bulbs remained dormant too many losses occurred through fungal and bacterial activity. Nevertheless, the dormancy of onions at harvest does allow such high temperatures to be used for the initial few weeks of storage to speed drying and enhance skin colour. If such high temperatures were used to enhance skin colour, at the end of the storage period the skin would still colour but the onions would sprout as their dormancy would have been broken by the period of storage at low temperature. Stored onions sprout for this reason as temperatures rise in the spring.

Spring flowering bulbs and corms enter a dormant phase in their native environment at the beginning of the hot dry summers. Although withholding water will frequently accelerate the development of dormancy in such species, the onset of dormancy seems to be under endogenous control or, at least, the environment plays a less important role that in many other species. During the period between the dying down of the leaves and the resumption of extension growth, spring flowering bulbs and corms are considered to be dormant. Very significant internal development occurs however in response to temperature changes.

These temperature-induced changes have been intensively studied by many workers, because control of temperature during the dormant phase can give horticulturists a considerable measure of control of flowering. Bulbs below a certain minimum size cannot be induced to flower. Commercial bulbs are sized using sieves with circular holes, the *circumferences* of which define the size grade of the bulbs. Bulbs of tulips have to be above 6 to 9 cm before they will flower and those of hyacinth 6 to 8 cm. For bulbs above the critical minimum size the general pattern of storage temperatures giving the earliest flowering of *Tulipa*, *Hyacinthus*, *Narcissus* and *Iris* species is high, followed by lower, followed by low. The timing of these phases and the actual optimal temperatures differ from species to species.

Flower initiation in *Narcissus* sp. starts before the foliage dies down and is often already completed before commercial stocks of bulbs are lifted in mid-summer. In contrast the apices of bulbs of *Tulipa* and *Hyacinthus* species are vegetative at this time and the whole process of flower initiation takes place during storage. Detailed programmes of storage temperatures have been developed to give almost any required control of the flowering process (Rees, 1966).

Onion sets are small bulbs produced by sowing seed in early summer at rates aimed to give about 1000 plants per square metre. They are stored through the winter and planted early the following spring, developing into larger and earlier-maturing bulbs than can usually be obtained from spring-sown seed. Certain cultivars are favoured because they give fewer flowering plants in the second year than others. If however the sets are large (more than about 15 mm in diameter), even these cultivars can give a high proportion of flowering plants. This tendency to flower can be eliminated by storing sets through the winter at about 25 to 30°C. Such sets are sold commercially as 'heat-treated'. It appears that prolonged high temperature prevents the flow primordia from forming; intermediate temperatures (8 to 12°C) are most favourable for flower bud initiation, and when stored at 0°C few sets subsequently bolt.

Potato tubers lifted at maturity exhibit differing degrees of dormancy depending on the cultivar and the conditions under which the crop has been grown. The tentative conclusion from recent experiments is that high levels of nitrogen fertiliser applied to the crop give tubers with a more prolonged dormancy.

A potato tuber (figure 4.4) is a specialised swollen stem, the eyes are axillary buds and the 'rose' end is the growing tip of the stem. At lifting

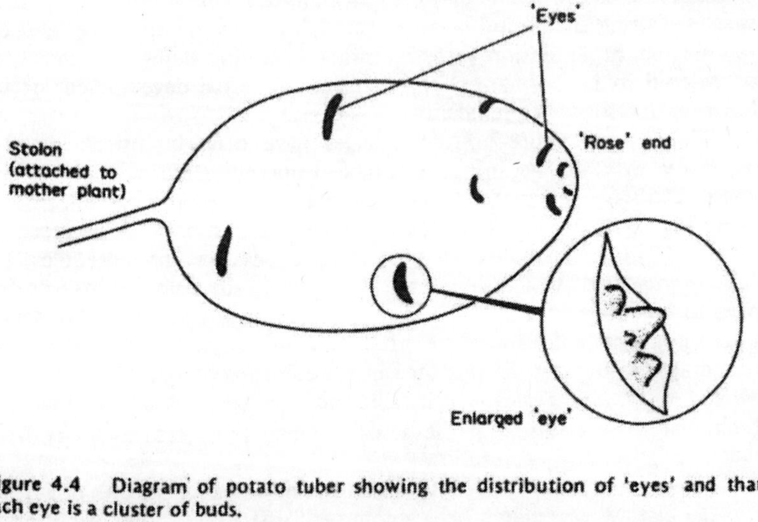

Figure 4.4 Diagram of potato tuber showing the distribution of 'eyes' and that each eye is a cluster of buds.

the buds at the rose end are the least dormant and can be forced into growth by about three weeks at 20°C. This 'opens' one or more of the apical eyes and progress of the treatment is easily monitored by periodic inspection. The heat-induced opening of an apical eye suppresses the growth of the other eyes on the tuber – it is said to exert apical dominance (see page 75). If as soon as one apical eye has started to grow the tubers are put in cool conditions (8°C), further growth can be restricted until such time as it is required, usually early in the following spring. Warmer conditions will then accelerate sprout growth but only the 'opened' eye will sprout, giving what is termed single-sprouted or apically-sprouted 'seed' tubers. These seed tubers give earlier yields because all the reserves of the mother tuber have been directed into a single sprout which makes rapid growth. If tubers are kept at a low temperature (5 to 10°C) from lifting, the dormancy of all or most of the eyes is removed and when temperature is raised in the spring a multi-sprouted 'seed' tuber is produced.

Each sprout from a 'seed' tuber can be regarded as giving rise to one potato plant. Thus if one batch of seed tubers is apically sprouted and another identical batch is multi-sprouted, and both are planted with similar spacing between the tubers, a higher population of stems (plants) will be obtained from the multi-sprouted seed. The crowding at the higher population will cause tubers to take longer to reach a usable size but will usually give a higher final yield.

The number of eyes on a potato is not directly proportional to its weight; larger potatoes have fewer eyes per unit weight. It follows that the amount of reserves per sprout can be increased by increasing the size of 'seed' tuber planted as well as by apical sprouting. If similar stem populations are established using large (60 g) and small (10 g) seed tubers, then the large ones give higher early yields but both give similar yields at maturity.

Also, if tubers are stored at temperatures below 10°C but above freezing their sugar content increases. Cooking the potatoes by frying in fat caramelises this sugar and imparts a dark colour and sometimes a bitter flavour. This sweetening can be reversed by putting the potatoes in a high temperature (18°C) for about two weeks before they are required.

CHEMICALS THAT BREAK DORMANCY

As previously indicated, the breaking of dormancy is frequency associated with the disappearance of inhibitor(s) rather than the appearance of a growth promoter. Gibberellic acid often plays a part in stimulating shoot growth once dormancy is broken, but it appears that this hormone can itself break the dormancy of some species. This applies particularly to woody species such as Horse Chestnut (*Aesculus hippocastanum*), Beech (*Fagus sylvatica*) and *Rhododendron* sp. Furthermore, kinetin causes bud break in some species, notably Apple (*Malus sylvestris*) and Vine (*Vitis vinifera*). Ether vapour or a solution of ethylene chlorhydrin will also break the dormancy of the buds of many woody species. The use of these chemicals for forcing early blooms of Lilac (*Syringa vulgaris*) or breaking the dormancy of Lily-of-the-Valley (*Convallaria majalis*) rhizomes is now well established.

Ethylene chlorhydrin (more correctly called 2-chloroethanol) can be used either as a vapour or a solution. (It is easier to use the vapour on trees and shrubs as soaking would be difficult.) The vapour or solution acts only on the buds actually treated, for if one bud of lilac is treated with vapour by sealing a test tube containing a few drops of 40 per cent ethylene chlorhydrin over it with modelling clay, then only that bud grows. The dormancy of the flower buds of *Rhododendron nudiflorum* can be overcome by exposing the plants for 24 hours to the vapour produced by a 40 per cent ethylene chlorhydrin solution at the rate of 7 ml per 100 litres.

Gladiolus corms of some cultivars can be successfully forced within a week of lifting provided they are treated for 3 to 4 days with ethylene chlorhydrin at the rate of 4 ml of 40 per cent solution per litre of corms.

Soaking in a 3 per cent solution of thiourea for 1 to 24 hours will release the eyes on a potato tuber from dormancy. Figure 4.5 shows how thiourea treatment causes all the buds within the eye to produce shoots, indicating that not only dormancy but also apical dominance are overcome by this treatment.

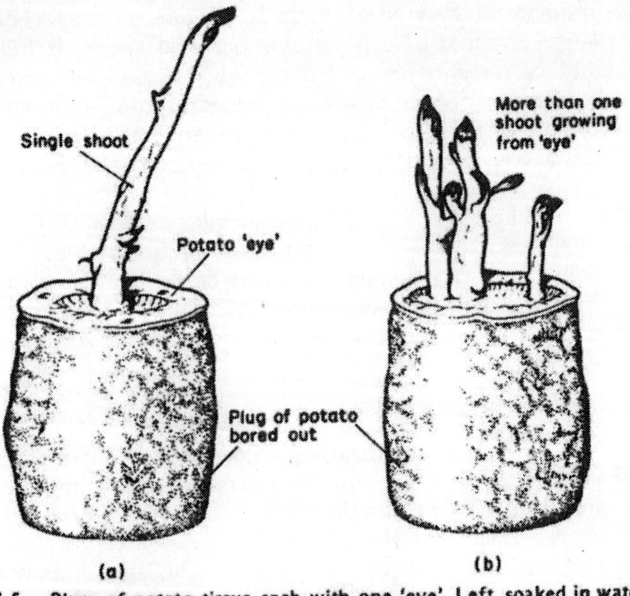

Figure 4.5 Plugs of potato tissue each with one 'eye'. Left, soaked in water; right, soaked in thiourea solution for 12 hours. The thiourea stimulates all the buds to grow.

Water is also a potent breaker of dormancy in some species. Dormant onion bulbs will begin to produce adventitious roots if they are wetted for 48 to 72 hours. Thus if it rains during the ripening of the crop in the field, bulbs are produced with a much reduced dormancy and sprouting in store becomes a serious problem. This happens even if the bulbs are subsequently dried and the roots shrivelled.

Immersion in hot water (30°C for 10 to 12 hours, or 40°C to 55°C for 15 seconds) is used commercially to break the dormancy of lilac and *Forsythia* flower buds. In this instance the correct temperature of the water is vital.

STERILE CULTURE AS A MEANS OF VEGETATIVE PROPAGATION

Individual cells or organs of many higher plants can be grown in a suitable nutrient solution by tissue and organ culture. The essentials for the success of these artificial cultures are to maintain the tissue or cells free from contamination by bacteria and to supply the complex organic compounds that would normally be derived from other cells or tissues in whole parts.

Plant physiologists have used tissue and cell culture to determine the role of many biological substances. Such studies have given rewarding insight into how hormones, vitamins and different proteins act.

Early attempts at tissue culture relied on using as nutrient sources the naturally occurring organic complexes of yeast extracts or coconut milk. These are still of value for the culture of material whose nutrient requirements are not fully known. However, for the last thirty years it has been possible to culture some plant tissues in fully synthetic media. Nutrient solutions typically contain the inorganic macro- and micro-elements (see page 40), sugar (either sucrose or glucose), vitamins (thiamin, nicotinic acid) and other organic chemicals, hormones (auxin, cytokinins) and organic complexes (yeast extract, coconut milk). These latter may be as much as 10 to 15 per cent by volume of the total solution. Such culture solutions can be made semi-solid by the addition of agar. This jelly-like medium is preferred for seed germination and embryo culture. Liquid media can be supplied to the tissue by a wick or the tissue can be continually bathed in nutrient by shaking or rotating the container. The isolated roots of some species will grow in static media but aeration or gentle shaking usually improves their growth.

The composition of the nutrient media, their use as gel or liquid, and the degree of shaking all combine to give the scientist a considerable degree of control over the culturing process. For example, transverse slices of carrot storage root will, in suitable conditions, give rise to callus in the region of the phloem. Small portions of this callus, which is an undifferentiated mass of largely parenchymatous cells, can be transferred to liquid culture. If the culture is agitated, cells at the surface break away and form free single cells. If these are put on to agar nutrient, each will

crop plants are now readily available as a result of tip culture; these crops include potatoes, apples, strawberries, carnations, chrysanthemums and, more recently, rhubarb.

The mass production of healthy stocks of genetically uniform plants is clearly possible for some species and it may not be too long before any species of plant can be multiplied using aseptic culture techniques. But this is not an unqualified blessing. Identical plants will succumb in an identical manner to pests, pathogens or environmental extremes. The variability in crops produced from seed usually avoids the complete loss of a crop, because some plants are able to survive adverse conditions. The growing of genetically *identical* crops is particularly risky for tree crops, which take many years to replace.

Nevertheless aseptic culture opens up exciting possibilities. Pollen cells can be cultured to give haploid plants, and if the chromosomes can be doubled by treatment with colchicine or radiation, then true-breeding diploid plants can be produced. It is also possible that new cultivars may be vegetatively bred by inducing two vegetative cells to fuse.

If we can culture the type of tissue we want, say apple flesh, why should we bother to grow apples on apple trees? Nature's way suddenly seems inefficient in the face of this surrealist effectiveness of vegetative propagation.

References

HANSEN, J., STRÖMQUIST, L.-H., and ERICSSON, A. (1978). Influence of the irradiance on carbohydrate content and rooting of cuttings of pine seedlings (*Pinus sylvestris* L.). *Plant Physiol., Lancaster*, 61, 975–79.

HESS, C. E. (1969). Internal and external factors regulating root initiation. In, *Root Growth* (Ed. W. J. Whittington). Butterworths, London, pp. 42–53.

KRAUSS, E. J., and KRAYBILL, H. R. (1918). Vegetation and reproduction with special reference to tomato. *Bull. Ore. agric. Exp. Stn.*, No. 149.

MOREL, G. M. (1964). Tissue culture – a new means of clonal propagation of orchids. *Bull. Am. Orchid Soc.*, June 1964, 473–8.

REES, A. R. (1966). The physiology of the ornamental bulbous plants. *Bot. Rev.*, 32, 1–23.

ROBERTSON, M., and WOOD, C. A. (1954). Studies in the development of strawberry II. Stolon production by first year plants in 1952. *J. hort. Sci.*, 29, 231–4.

THOMAS, T. H. (1969). The role of growth substances in the regulation of onion bulb dormancy. *J. exp. Bot.*, 20, 124–37.

WENT, F. W., and THIMANN, K. V. (1937). *Phytohormones*. Macmillan, New York.

WINKLER, A. J. (1927). Some factors influencing the rooting of vine cuttings. *Hilgardia*, 2, 329–49.

Further Reading

HARTMANN. H. T., and KESTER, D. E. (1968). *Plant Propagation Principles and Practices*. 2nd Ed. Prentice-Hall, London.

IDLE, D. B. (1968). Ice in plants. *Science J.*, 4 (1), 59–63.

MARYLAND, H. F., and CARY, J. W. (1970). Frost and chilling injury to growing plants. *Adv. Agron*., 22, 203–34.

PEARSE, H. L. (1948). Growth substances and their practical importance in horticulture. *Tech. Commun. Bur. Hort. Plantr. Crops*, No. 20.

CHAPTER 5

THE FLOWER

Sooner or later the vegetative higher plant undergoes a profound change and flowers. Even casual observation will reveal that many species have sharply defined seasons for flowering. Some seasonal change induces the plant to turn from the vegetative to the flowering state. Other species flower at any season that permits growth, but it seems that they must attain some minimum amount of vegetative growth before flower buds can be produced. This period of vegetative growth may last several years in woody species or be only a few weeks in short-lived annuals.

The apex of the vegetative plant somehow is induced to form flower buds instead of stem and leaves. Physiologists studying this change at the cellular level recognise that one of the earliest signs that flowering has been induced is the activation of mitosis in the meristem flanks. Mitotic activation progresses upwards until the whole apex is affected. In general, this alters the shape of the apex, seemingly because different parts respond differently to induction. There are also very considerable alterations of the metabolism of the apical region; these changes are undoubtedly significant in flowering but we do not yet understand how they relate to the morphological changes that can be seen. How do genes determine the ultimate form of an organism? What are the details of the control system that ensures that seeds which are very similar in appearance (say cabbage, cauliflower and Brussels sprout) develop each according to its kind? The transition of an apex from the vegetative to a flowering state is a profound morphogenetic event which lends itself to studies aimed at answering these questions, at least in part.

Plant physiologists, like other scientists, find that the ability to control a process is a key to the understanding of the process. Finding the key will, as it were, tell us something about the lock. Continuing the analogy, horticulturists are likely to be most interested in knowing about the different kinds of keys that can unlock flowering and will be less concerned with the locks. Accordingly this chapter is primarily about the control of flowering.

RIPENESS-TO-FLOWER

In many species external stimuli that promote the flowering of older plants do not induce flowering in young plants. These young plants are thought of as being in a juvenile phase, and are said to have reached *puberty* when they are receptive to flowering stimuli.

The existence of a juvenile phase is of considerable horticultural importance. We shall see that many *Brassica* species, such as cabbage and Brussels sprout, require a cold stimulus for flowering, but the young seedlings can be subjected to prolonged periods at low temperature without bolting being induced. Thus, in temperate regions small plants can be grown from seed just before winter sets in, kept through the winter with some protection if necessary and be planted out in the spring. If, however, the plants are too big and have passed their juvenile phase before winter starts, then the period at low temperature will cause subsequent bolting. Cultivars within a species differ in the 'size' at which they reach their puberty; those with a particularly advanced puberty are more suitable for over-wintering as seedlings.

Within a cultivar puberty tends to be reached at a constant physiological age rather than after a constant time from sowing. The concept of physiological age is a most important one in studies of flowering. For example, if we were to study the effect of temperature on flowering, we would find that sunflowers (*Helianthus annuus*) flower sooner at a high temperature (25°C) than at a low one (10°C). We might conclude that flowering in sunflowers was favoured by high temperature. This would be wrong, because the higher temperature favours growth and not flowering. To separate the effects of a given treatment on growth from those on development, we need to compare the flowering of the plants on some other basis than astronomical time. Physiologists usually count the number of leaves on the main stem as a developmental time scale. Thus, if our two sunflowers each produced a flower after twenty leaves had formed, we would conclude that temperature had no effect on flowering. This leaf-number time scale is sometimes referred to as a *plastochrone* scale. (More strictly, a plastochrone is the time between the initiation of two successive leaves on a stem.)

Two cultivars of a species may differ in their rates of growth in the same environment, but not in their rates of development: both would flower after twenty leaves, but the slower grower would take longer to reach the twenty leaf stage. Conversely they may both flower at the same *time* but at different stages of development (say twenty and twenty-five leaf stages). The concept of physiological age allows us to differentiate between effects on growth and development.

'NEUTRAL' SPECIES

Flowering plants which do not respond to day length by changing their rate of flower development are classed as 'neutral'. The sunflower, broad bean (*Vicia faba*) and tomato (*Lycopersicum esculentum*) are amongst the host of higher plants in this category.

In these species there is a strong internal regulation of flowering which imposes a vegetative phase before flowering occurs. Then there is a transition phase in which the apex changes from its markedly vegetative form to the flowering form. This progressive change in the apex is often accompanied by other progressive changes (for instance, in leaf shape). Leaves of a more complex shape are produced as the apex approaches the flowering state. This can be seen by examining plants of the common weed, groundsel (*Senecio vulgaris*), which is another 'neutral' species.

PHOTOPERIODISM AND THE INDUCTION OF FLOWERING

Many higher plants flower in response to the duration of the period of light within the daily cycle of 24 hours.

Short-day plants will only flower when the light period is shorter than some critical period.

Long-day plants will only flower when the light period is longer than some critical period.

These definitions indicate that there is no absolute duration for a short or a long light period. Some long- and some short-day species can be induced to flower in a 12 hour light period but they will differ in their response when the light period is lengthened or shortened. Once flowering has been induced in most species, the photoperiod can be altered a great deal and persist at some unfavourable level for a long time before the plant will revert to its vegetative state. The showy annual Cosmos (*C. bipinnatus*) provides a good example of the persistence of the response to the flowering stimulus. In temperate regions this plant will not flower if sown in the long days of summer (late May in UK) but if sown earlier or if the day length is reduced by covering the seedlings, then flowering will be induced by the period in short days and persist throughout the long days of summer. With some cultivars of this species 10 short-days, each of less than 10 hours, are sufficient to induce flowering. This treatment can be horticulturally worthwhile as it avoids the growing of barren giants that only flower just before they are killed by frost.

Some species, such as chrysanthemum and strawberry (*Fragaria chiloensis*), which must have short days to flower are termed *absolute* or *qualitative short-day species.* There are also absolute or qualitative long-day species, such as spinach (*Spinacia oleracea*) and *Sedum spectabilis.*

Other species will flower in any day length but are not neutral species because flowering is advanced in some of them by short days and in others by long days. These quantitative species include Cosmos, cotton (*Gossypium hirsutum*) and rice (*Oriza sativa*) in the short-day group, and snapdragons (*Antirrhinum majus*), petunia (*Petunia hybrida*) and the garden pea (*Pisum sativum*) in the quantitative long-day group.

Even these groupings do not cover all the possible types of flowering response to photoperiod. Some species require to be exposed to short days before they will flower in response to long days. The common clover

(*Trifolium repens*) is such a short–long-day species and *Bryophyllum crenatum* is a long–short-day species. Some species, including some cultivars of sugar cane (*Saccharum officinarum*), only flower in intermediate day lengths, others do not flower at intermediate day lengths but will in both longer or shorter ones. This phenomenon is termed *ambiphotoperiodism*.

Physiologists, not surprisingly, are mainly interested in species that are particularly responsive to photoperiod and/or are amenable to some technique, particularly grafting. Among the classical species for study are

Xanthium pennsylvanicum (Cocklebur), which has an absolute requirement for short days and is very responsive; a flowering apex is readily discernable at an early stage.

Perilla ocymoides is an absolute short-day species particularly favoured for experiments involving leaf grafting.

Kalanchoë blossfeldiana is an absolute short-day species which produces thick leaves in short days and thinner ones in long days.

Pharbitis nil (Japanese morning glory) is an absolute short-day species which will respond even at the cotyledonary stage of growth.

Hyoscyamus niger (Henbane) is an absolute long-day species, not responsive until it is some 10 to 30 days old.

Glycine max (Soybean) is an absolute short-day species, used in classical experiments in which 'days' of more than 24 hours were used and in studies of the interaction between temperature and photoperiod, etc.

Lolium temulentum (Darnel rye grass sp.) is an absolute long-day species responding at very early stages of growth.

Studies with these and other species have revealed a great deal about photoperiodism. For example, it is wrong to classify the species according to the way they respond to periods of light (although it is always done), because it is the duration of the dark period which is significant in the flowering stimulus. Typical of many experiments on this point are those in which the length of the light/dark period is reduced from 24 hours to 16 hours. A short-day plant will flower if the light period is 8 hours and the dark period 16 hours, but if the light period is 8 hours and the dark is also 8 hours it will not flower. Furthermore, if every other light-period of 8 hours is shortened to a few minutes then we are getting near to the 8 hours light and 16 hours darkness which induced flowering. But in *Xanthium* the gap between the two 8 hour dark periods needs only to be 1 minute to prevent flowering.

This clearly demonstrates that the length of the dark period is important in short-day plants – they are really long-night plants. A short dark period in the middle of a long day will not stop the flowering of long-day plants or cause short-day plants to flower. If long-day plants are in short days, however, and the night is interrupted at near its mid-point, flowering will occur. *Hyoscyamus* responds to light-breaks of 0.1 to 1 hour.

Night-break treatments

Night-break treatments have become of considerable significance in growing chrysanthemums. Night-breaks keep plants in their vegetative state when the natural day length is short. The commercial techniques used in producing chrysanthemums throughout the year will be outlined later in this chapter, after some of the principles involved have been more fully described.

High intensity light is needed for photosynthesis and certain regions of the visible spectrum are more efficient energy sources than others (see chapter 2). Photoperiodic effects are brought about by light of very low intensity and, as other regions of the spectrum are involved, a pigment (or pigments) other than chlorophyll is the receptor.

Phytochrome is the name given to the pigment associated with the absorption of light that causes morphogenetic responses. Borthwick and Hendricks investigated the action spectrum of the night-break treatments of a short-day plant and also the light requirement for the germination of lettuce seed. By the early 1950s they and their co-workers had established that both these effects were mediated by the same pigment system.

The most effective wavelength region of the spectrum for night-break treatment of short-day plants is 600 to 650 nm (the orange-red region of the visible part of the spectrum). Light of this wavelength region promotes lettuce seed germination but far-red (700 to 800 nm) inhibits germination, even if red light is given previously; inhibition is removed if red light is applied subsequently. The seed therefore responds to that wavelength of light it receives last of all. This is also true for night-break treatments of short-day plants. The effect of a red light night-break can be completely nullified if followed by a period of far-red radiation.

These and other observations led to the suggestion that a single pigment existing in two interconvertible forms was involved.

$$P_r \underset{\text{red}}{\overset{\text{far-red}}{\rightleftharpoons}} P_{fr}$$

The P_r form of phytochrome absorbs red light to become the P_{fr} form which absorbs far-red radiation. The extraction, purification and assay of this pigment system has confirmed this theory (Siegelman and Firer, 1964).

This phytochrome system appears to be an important element of the 'clock' which measures photoperiodic 'time'. In daylight the P_r and P_{fr} forms of the pigment exist in about equal proportions, but in darkness P_{fr} is slowly changed to P_r. Thus if we wish to redress this drift to P_r to obtain the daylight situation, red light is needed. This reverses the trend and the new balance of the phytochrome forms misleads the plant into behaving as though it has been in long days. If the plant is of a short-day species flowering is inhibited, if it is a long-day plant flowering is induced.

There are many exceptions to this simplified view of the role of phytochrome in photoperiodic perception. The exception is more likely to provide the clue which will enable a more all-embracing hypothesis to be formulated and tested. Nevertheless, it is generally accepted that the phytochrome system somehow plays a major part in the photoperiodic control of flowering. This tells us something of the quality of light we need for photoperiodic effects; red light is the most effective for night-break treatment. Fortunately red light is a major component of the light from ordinary tungsten filament bulbs but it has been shown (Canham, 1966) that red magnesium arsenate and warm white fluorescent tubes can be even more efficient for night-break treatments (figure 5.1).

When a daylight day is extended artificially to produce a long-day effect, it is important to supply a proportion of far-red light, so tungsten filament bulbs are more suitable. Very low intensities can be effective if supplied over long periods; bright moonlight (at about 0.02 lux) is

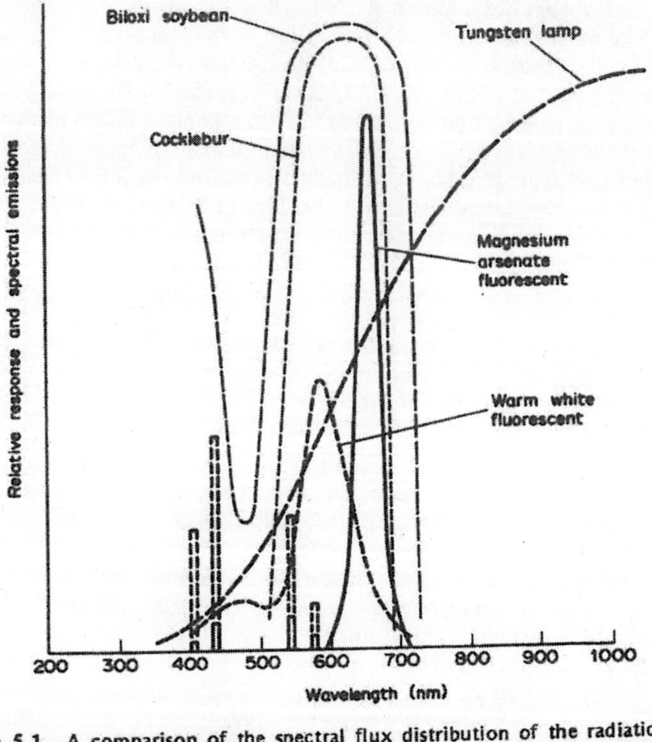

Figure 5.1 A comparison of the spectral flux distribution of the radiation from three types of lamp with the night-break action spectrum of Biloxi soybean and cocklebur.

however about ten times too low to produce photoperiodic effects. For successful night-break treatment the light intensity at plant level should not be lower than 100 to 200 lux.

As we shall see later, night-break treatments are used to produce chrysanthemums throughout the year by prolonging vegetative growth. The flowering of antirrhinums and many fuchsia cultivars can also be promoted in this way. Shortening the natural daylight period by covering plants is used commercially to promote the flowering of chrysanthemum and poinsettia (*Euphorbia pulcherrima*).

The receptors of photoperiod stimuli

Long before the discovery of phytochrome physiologists tried to determine which organs were perceiving the light stimulus for flowering. Garner and Allard were the first to establish beyond any doubt that photoperiod controlled flowering in some species. Chailakhyan in Russia first showed that the leaves perceive the flowering stimulus. He suggested that induced leaves produce a hormone, which he named **florigen**. Very young and old leaves are less effective producers of florigen than medium-aged leaves. Florigen can be graft transmitted but will not impart inductive powers to other leaves. Only those leaves receiving the day-length treatment are induced to form florigen – other leaves on the same plant not receiving inductive light periods do not form florigen. These generalised statements are summarised diagrammatically in figure 5.2.

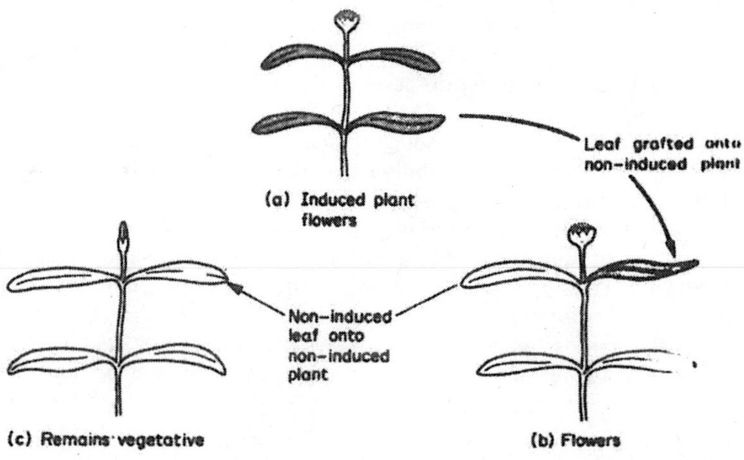

Figure 5.2 Grafting experiments similar to the one represented here have shown that the flowering stimulus is only produced by leaves that have been in an inductive photoperiod.

As usual, generalisations of this kind have exceptions: in some species the grafting of a leaf from plant (b) in figure 5.2 induces flowering in plant (c), and these leaves are said to have been secondarily induced.

The flowering-hormone is not transmitted across a gap between cut surfaces held apart by a gel or some other permeable barrier that prohibits a true graft union. There is some evidence for the existence of an anti-flowering hormone, in that non-induced leaves can apparently inhibit flowering. Plant (b) in figure 5.2 would be more likely to flower if the non-induced leaves were removed.

For decades plant physiologists have failed to isolate the flowering hormone but an interplay of hormones is most likely the cause of flower induction.

HORMONES AND FLOWER INDUCTION

From what has been indicated so far, it will be apparent that if florigen is responsible for flowering, the hormone must be formed during short days in some species and during long days in others. Several experiments with inter-specific grafts indicate that this may be so. For example, the short-day species *Xanthium* will flower in long days if grafted on to the long-day species *Rudbeckia*.

Several known plant hormones seem to play some part in inducing flower initiation. The gibberellins will promote flowering in several long-day species but not in short-day ones. They will also induce the flowering of very young seedlings of some coniferous trees (for example *Cupressus arizonica*). On the other hand, some growth retardants that are regarded as anti-gibberellins will also promote flowering in species of *Rhododendron*. The level of gibberellins seems therefore to be critical for flowering in some species; the level in the vegetative state may be either too high or too low for flowering, depending on the species.

Auxin applications will induce flowering in a few species notably the pineapple, for which the technique is used commercially. Just as with gibberellins the concentration within the plant seems to be vital, since auxin application can inhibit as well as induce flowering.

Just how the day length induces flower-bud formation and other changes in plants remains largely unknown. Hormones, some known and perhaps some not yet discovered, are involved and phytochrome recognises the light periods encountered. Perhaps we would understand flower induction better if the flowering in the apparently simpler day-neutral species had been more studied.

VERNALISATION

The breaking of seed dormancy and the induction of flowering by cold treatments (known as vernalisation) were described in chapter 1. The winter-annual strain of Petkus rye, which can receive its cold stimulus as an imbibed seed, differs in its behaviour from that of biennials in that the latter must reach a certain size or degree of maturity before they will respond to low temperature and flower (see page 15). Brussels sprouts and cabbage come in this category. Other biennials can receive at least some of their low-temperature requirement before the seed germinates or very shortly after emergence. In beet (*Beta vulgaris*) recent work has shown that low temperatures during the seed ripening phase can reduce the low-temperature requirement for flowering of plants.

A low temperature favours seed-stalk development in onions. Hence over-wintered seedlings are inclined to bolt in the following year but, because a certain minimum size has to be reached before the plants respond to cold, small plants can be over-wintered without subsequent bolting.

The vernalisation of plants is a quantitative and reversible process; temperature and time of vernalisation are related, but not in a simple way. Although the lower the temperature the shorter usually is the period required for induction, temperatures near freezing (<2°C) are in general less effective than slightly higher temperatures (2 to 5°C). In some species or cultivars quite high temperatures are recognised as being low. This applies to curd initiation in cauliflowers, which will recognise temperatures of about 15°C as low, according to cultivar.

Vernalisation is peculiar in that it appears to involve chemical changes that are favoured by lower temperatures. (Chemical reactions involving enzymes are usually favoured by higher temperatures.) This can be accounted for by assuming that two enzymes (A and B) are competing for the same substrate C and, although the two enzymes are both more active as temperature rises, they differ because A is better than B at low temperature whereas B is better than A at high temperature. There is no experimental evidence to support the view that such a system operates during vernalisation but it does go some way to explaining the anomaly.

The examples of the flowering stimulus produced by vernalisation being transmitted by grafting seem to be exceptions to a general rule. Only cells derived from a chilled region of dividing cells will normally behave as though vernalised. Thus, if the tip of an unvernalised plant is grafted on to a vernalised stock, there will be no flowering; a vernalised scion would flower on an unvernalised stock however.

The growing tip of a plant is normally the site sensitive to vernalisation. Experiments with small cooling coils have shown that if the tip is cooled all subsequent tissues developed from that tip are vernalised. There is evidence in *Lunaria biennis* that any group of dividing cells (not necessarily a shoot tip) can be vernalised. This species is day-neutral but will not flower unless vernalised. Leaf cuttings will root and give adventitious shoots, which will flower if the detached leaf has first been held for a period at 5°C. However if the 5 mm of petiole nearest the cut end is removed after the cold treatment, the subsequently formed shoots do not flower. This indicates that the dividing cells near the cut end are sensitive to the cold and not the leaf as a whole.

THE INTERACTION BETWEEN VERNALISATION AND PHOTOPERIOD

Many biennial species in which flower formation is induced by vernalisation need long days for further development of the seed stalk. In red beet the growth of the seed stalk is delayed or even prevented if the photoperiod is less than 12 hours, whereas photoperiods in excess of 14 hours will hasten the growth.

To illustrate how vernalisation and photoperiod regulate flowering in some species, details of some recent work with cabbage will be cited and a general account of the growth and flowering of chrysanthemum given.

Cabbage

Heide (1970) studied four cultivars of early round-headed cabbage which behaved similarly. The time between the cessation of the cold treatment and flowering (anthesis) increased as the temperature during the cold treatment increased. At 4°C the period was 56 days and at 7°C and 10°C, 58 and 66 days, respectively. Increasing the duration of the cold treatment reduced the time to flowering from 85 days for 4 weeks' exposure to 63 and 50 days for 6 and 8 weeks' respectively.

Figure 5.3 summarises the results of an experiment where a standard chilling temperature of 5°C was supplied for three periods to plants aged between 2 and 12 weeks. It can be seen that sensitivity to cold, in terms of the number of bolters, increased with plant age up to an age of 5 to 6 weeks. At this age 60 per cent of the plants bolted after as little as three weeks at 5°C whereas younger plants needed 6 to 9 weeks' chilling to reach this figure. There was some indication that 12 week old plants were less sensitive to 3 weeks' chilling than younger plants, but in general maximum sensitivity to chilling had been reached by an age of 5 to 6 weeks.

When the chilling period was not continuous but was interspersed with periods at a higher temperature, complete devernalisation occurred if 16 hours of each day were at 5°C and the remaining 8 hours were at either 18°C or 24°C. Partial devernalisation, expressed as delayed flowering, occurred when the 8 hour period was at 12°C. From this and other experiments, Heide concluded that any temperature that is too high to vernalise cabbages actively devernalises them – there is no neutral temperature.

In all the above experiments Heide continuously illuminated the plants during the chilling treatment and then gave 15 hour photoperiods. In another experiment the effects of short (8 hour) and long (24 hour) days, both during exposure to low temperature (5°C for 6 weeks) and

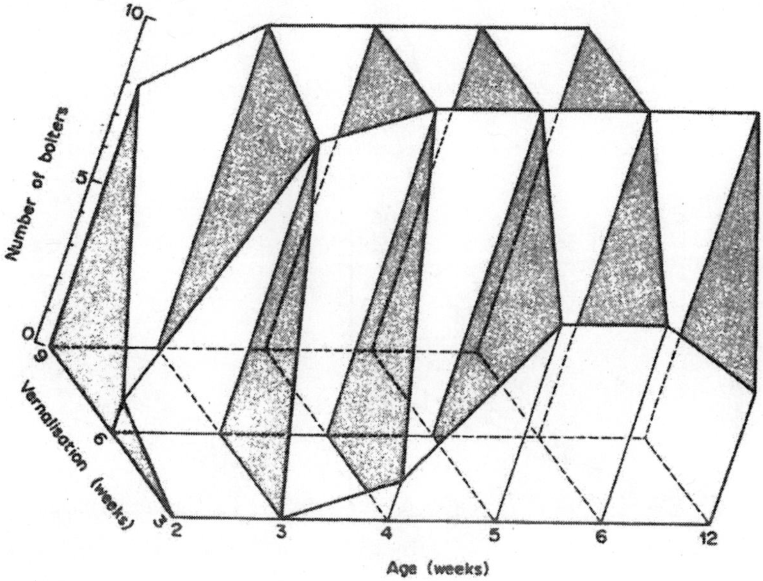

Figure 5.3 The subsequent bolting of cabbage plants chilled (5°C) at various ages for 3, 6 or 9 weeks (from Heide, 1970).

subsequently, were examined using a range of temperature for the post-chilling period. The results Heide obtained are summarised in figure 5.4. Long days during chilling enhanced stem elongation but did not markedly reduce the number of days to anthesis (numbers underlined in figure 5.4). Long days following chilling considerably enhanced stem elongation and advanced flowering however. These effects of photoperiod on stem elongation following chilling only occurred at temperatures below 18°C.

These and other results may be summarised as follows

(1) Temperatures below 12°C are effective in producing flowering but the optimum temperature is in the range 4 to 7°C.
(2) At the optimum temperature (5°C) only 3 to 4 weeks' chilling is needed, whereas 12°C has to be maintained for 6 months to induce flowering.
(3) Sensitivity to chilling increases with age up to 5 to 6 weeks, when the plants have 7 to 9 leaves.
(4) Expression of the vernalised state depends on the subsequent temperature and photoperiod.

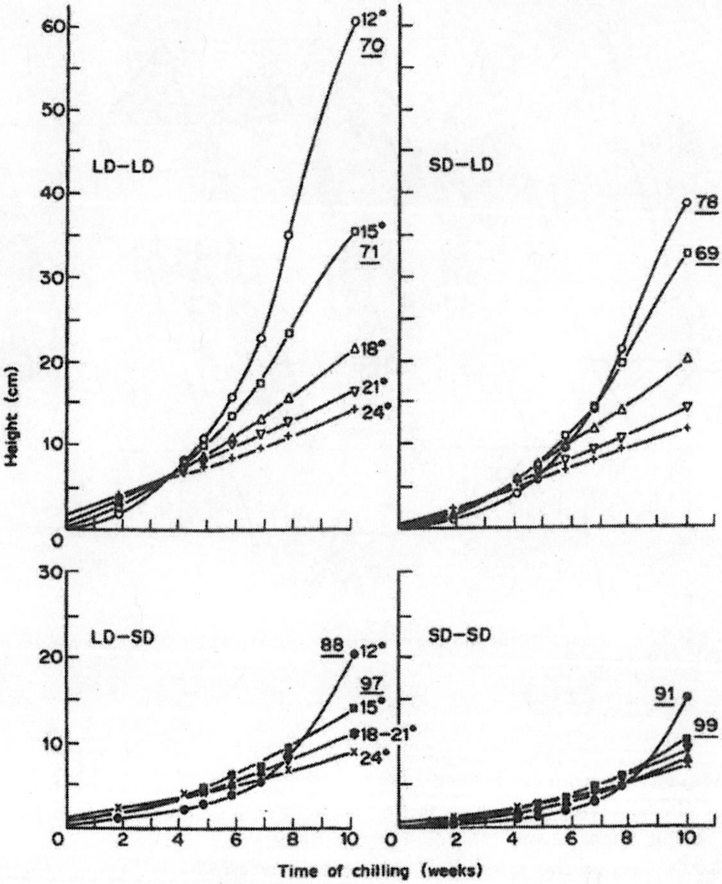

Figure 5.4 The course of stem elongation in cabbage plants grown at different temperatures and day lengths following exposure to 5°C in 24 hour (LD) or 8 hour (SD) photoperiods and subsequently grown at the temperatures indicated. Underlined figures are days to anthesis (from Heide, 1970).

(5) The devernalising effect produced by subsequent high temperatures increases with temperature and duration, but there is no devernalisation effect if the plants receive more than 6 weeks' chilling at the optimum temperature.

Chrysanthemums

Although many of the so-called 'early' cultivars show little or no flowering response to photoperiods, artificial light has been used for more than fifty years to delay the flowering of chrysanthemums. Initially this light was added to the end of the natural period of daylight to lengthen the photoperiod, but in the 1940s it was shown that a period of light near the middle of the night was just as effective. Post (1953) showed that 250 lux hours each night-break was all that was needed to prevent flowering in some cultivars of glasshouse chrysanthemum. Furthermore, 500 lux for half an hour was just as effective as 50 lux for 5 hours.

Borthwick and Cathey (1962) demonstrated that the effectiveness of the night-break treatment depends on the conversion of phytochrome to the far-red-absorbing form (P_{fr}), which can be maintained at an effective level by intermittent lighting. The dark periods between successive periods of light must be no longer than 30 minutes if the light comes from incandescent lamps. The length of the permissible dark period depends on the intensity and wavelength of the light applied during the light period. The higher the red:far red ratio and the higher the intensity, the more effective was the conversion $P_r \xrightarrow{\text{red}} P_{fr}$. The more P_{fr} formed, the longer could be the subsequent dark period, during which P_{fr} reverts to P_r.

Incandescent lamps giving an intensity of 200 lux need to be lit for only 5 per cent of a cycle, provided the cycle is less than 30 minutes and used for a 4 hour period during the mid-period of each night. Thus, 3 seconds of light every minute for 4 hours is as effective at preventing flowering as 1.5 minutes of light every 30 minutes for 4 hours or even continuous light for 4 hours.

The same quantitative relationship between intensity and duration established by Post (see above) for conventional night-break treatments also applies to these cyclic night breaks. Thus if light of only 50 lux intensity is used, illumination must be applied for 20 per cent of each cycle. Cycle times of less than 10 minutes tend to produce excessive wear on switch gear and reduce lamp life and consequently are not recommended for commercial use. The minimum practicable cycle would be 30 seconds at 200 lux in every 10 minutes.

The use of cyclic night-break lighting can save 80 per cent, or more, of the power cost of conventional night-break lighting systems, but this has to be set against the increased capital outlay for switches and time clocks. The period and duration for which night-break or cyclic night-break treatments are applied depends on (1) the cultivar (naturally dwarf cultivars may need a longer period of vegetative growth than naturally tall ones), (2) the intended use (more vegetative growth being required for cut blooms than for plants sold in pots), and (3) the natural day length (the shorter the natural day the longer the night-break period). Short days induce flowering and in commercial practice these immediately follow any period of long days that may have been needed to establish the required vegetative framework. Provided that the natural day length is below 12 hours, most cultivars will not need any artificial shortening of the day. When the period of natural light exceeds 12 hours, however, covers made of black polyethylene or closely woven cloth are used to exclude daylight for in excess of 12 hours each day. Any tears or pin pricks allowing light to enter will cause uneven flowering. If the covers are used for only 6 days each week, there is usually a 2 to 3 day delay in flowering. The effectiveness of shortening the days depends in some cultivars on the temperature not falling below 16°C during the induction of flowers.

Many of the cultivars of chrysanthemum must be vernalised before they will respond to day length. In the unvernalised state the shoots have very short internodes and are diageotropic. This is typically the state of a plant that has flowered and is producing new basal shoots that can be used for cuttings. About 3 to 4 weeks of chilling are needed for full vernalisation. The perennial nature of chrysanthemums seems to depend on devernalisation caused by the low light intensities and high temperatures occurring at the base of the plant in the warm days of late summer. These basal shoots are devernalised during their formation, only to be vernalised by the winter cold and so rendered sensitive to the short days of the late summer. They are not sensitive to the short days of spring because vernalisation is not complete at this time.

It appears that devernalisation in chrysanthemum may be typical of that in many perennial plants. It is interesting that low light intensities are essential. Studies of the mechanism of this devernalisation indicate that increased auxin and decreased gibberellin levels are involved.

TEMPERATURE AND PHOTOPERIOD

Apart from vernalisation, the temperature at which plants are grown can modify the effect of a given photoperiod. The cultivar of tobacco, Maryland Mammoth, on which Garner and Allard originally established the phenomenon of photoperiodism, illustrates the modifying effect of temperature. At 18°C plants of Maryland Mammoth will flower in 9 to 10 hour photoperiods but remain vegetative in 16 to 18 hour photoperiods. If the night temperature is lowered to 13°C flowering occurs in both these photoperiods. Flower induction in both long- and short-day plants is influenced more by the night temperature than the day temperature. There are no hard and fast rules but in general high night temperatures reinforce short-day effects and low night temperatures long-day effects.

Annuals, winter-annuals, biennials and perennials

Flowering plants can be divided into two major groups: *monocarpic* (which flower once and then die) and *polycarpic* (which do not die after flowering but survive to produce successive crops of flowers). Annuals are monocarpic and may or may not be photoperiodic, but they do not require a cold stimulus. Examples are groundsel (*Senecio vulgaris*), spring wheat and tobacco.

Winter-annuals are monocarpic, receptive to a cold stimulus as imbibed seed and may or may not be photoperiodic. Examples are *Aira praecox* and *Myosotis discolor*.

Biennials are monocarpic and, like winter-annuals, require vernalisation, but they are not as sensitive as seed or as young seedlings as they are once a 'juvenile' phase has passed. Thus, they tend to be vegetative in their first year and flower in their second year. They may be day-neutral or have a short-day, or a long-day response. Examples are cabbage, celery and foxglove (*Digitalis purpurea*).

Perennials are polycarpic and somewhat similar to biennials, but the plant survives flowering and needs to be revernalised before it will flower again. Examples are primrose (*Primula vulgaris*), Michaelmas daisy (*Aster* sp.) and rye-grass (*Lolium perenne*).

FLOWERING STIMULI AS DETERMINANTS OF SPECIES DISTRIBUTION

If a species is to survive, it must reproduce to make good the wastage caused by natural disaster. Some species with highly effective methods of vegetative reproduction (for instance, *Lemna minor*, see page 51) have no need of a seed-phase to ensure this survival. The evolutionary success of flowering plants depends largely on the ability of their seed to survive adverse conditions. Flowering is therefore an essential for the survival of many plants. These plants must grow in the environment that provides the necessary stimuli for flowering. Long-day species of temperate zones will not flower in the short days of the tropics. The chilling requirements of winter-annuals, biennials and perennials are also not provided by tropical environments.

Wild species with a wide distribution often exist as a series of races, each adapted for flowering to the environment of a limited range of latitudes. For example, *Bouteloua curtispendula* (side-oats grama) is a native of America and extends northwards into Canada and southwards into Mexico. Twelve strains of this species collected within an area covered by 17 degrees of latitude showed different photoperiodic responses. Plants of the most southern strains failed to flower in photoperiods longer than 14 hours, whereas northern strains flowered normally in such photoperiods.

POLLINATION

The formation and growth of flower buds ultimately gives the opened flower with receptive stigma awaiting pollination. Pollen from the same flower or from a flower of the same or different species may be transferred to the stigma by wind, insect or water.

Germination of pollen – evidenced by the growth of a pollen tube – will usually occur within a few minutes of arriving on a mature stigma. The growth of a tube may be slower than that of other tubes from pollen grains on the same stigma. This may be caused by a lack of compatibility between the pollen tube and the tissues of the pistil. The developing pollen tube obtains water and inorganic and organic substances from the tissues of the pistil. Incompatible pollen apparently is unable to parasitise the pistillar tissue on which it finds itself; it may be unable to produce the right enzymes or find the organic nutrients that it needs.

Incompatibility can occur between the pollen and stigma of the same plant. This type of self-incompatibility prevents self-pollination, but it is usually not of an all or nothing effect. A plant's own pollen will affect a limited fertilisation, provided there is no competition from compatible foreign pollen.

Although pollen from several different plants may all be present on a stigma, the grain whose tube reaches the embryo sac first and completes fertilisation is the only one that matters. This success is largely determined by compatibility but other physiological factors are also important. The unsuccessful pollen may be too old; the pollen of grasses is short-lived and perishes within a few hours. Other pollen, in particular that of trees, is much more long-lived and when stored at low humidities (10 to 50 per cent RH) and low temperatures (0 to 10°C), it can be collected one year and used the next.

Vacuum drying and freeze drying have also proved successful; oil palm (*Elacis guineensis*) pollen only remains viable for 1 week under natural conditions, but pollen that has been vacuum-dried for 1 hour and stored under vacuum in ampoules at 28°C, 5°C or −10°C loses little of its viability, even after a year (Hardon and Davies, 1969).

Reasonable longevity can be a considerable aid to breeding programmes, because it removes the need to synchronise the flowering of strains that are to be cross-bred. Breeders can exchange pollen of promising lines without giving away parent material. Thus, the potential of a hybrid combination can be assessed even although the parents may belong to two rival seed companies. If the offspring is sufficiently attractive rivalry can become a partnership to everyone's advantage.

If the temperature is too low, pollen may fail to germinate on a stigma. The relationship between pollen-tube growth and temperature is a characteristic of a species. The relationship that is established with the pollen in some artificial medium, such as sugar solution, may not be the same as that in more natural circumstances. Using detached flowers of tomato, it was found that germination and fertilisation is most rapid at the highest temperature (37.5°C), but at this temperature less than one-eighth of the pollen germinates when compared with that at the optimum temperature for germination (25°C). Germination at 5°C takes 20 hours (as opposed to 1.3 hours at 20°C and 0.5 hours at 37.5°C), and pollen-tube growth is so slow that pollination, which takes less than 1 hour at 37.5°C, was calculated to have taken about 6 to 7 days.

To be successful, the pollen grain not only has to produce a fast growing tube, but this tube must also grow in the direction of the embryo sac. Molisch, working in the 1890s, showed that pollen tubes grew along gradients of increasing concentrations of chemicals diffusing from pistillar tissue. It appears that these chemotropically active substances are not specific for the plant that produces them because pollens of many different species respond similarly. Recent work has suggested that a calcium gradient may be responsible for guiding pollen-tube growth in several different species.

MALE AND FEMALE FLOWERS

Although most species of flowering plants produce bisexual flowers, some species are dioecious – with male and female flowers on different plants (for example, spinach); and others are monoecious – with the separate male and female flowers on the same plant. Many cucurbits are monoecious and because only the female flowers produce fruit, the proper ratio of male to female flowers is of horticultural significance.

In cucumber (*Cucumis sativa*) or squash (*Cucurbita pepo*) all the first formed flowers are usually male. Then a mixture of male and female flowers is produced, with an ever increasing proportion of female flowers for the rest of the plant's life. Spraying auxin on to the leaves of young plants accelerates this progression towards femaleness, but spraying gibberellins has the opposite effect. Some cultivars of cucumber that normally only give female flowers can be induced to produce male flowers by sprays of gibberellic acid. Many experiments have indicated that maleness is favoured by short days and femaleness by long days.

In some species that normally produce perfect flowers, individual plants can be found with incomplete, and therefore ineffective, development of one of the sex organs. Forms with certain types of inherited male sterility are of considerable value to plant breeders for the production of new uniform hybrid cultivars. Some forms of male sterility in tomato and barley can be overcome by spraying the young flower buds with gibberellic acid. It may be useful if male sterility can be so reversed in crops where it is not otherwise available in a convenient form for the plant breeder.

References

BORTHWICK, H. A., and CATHEY, H. M. (1962). The role of phytochrome in control of flowering chrysanthemum. *Bot. Gaz.*, 123, 155–62.

BORTHWICK, H. A., HENDRICKS, S. B., PARKER, M. W., TOOLE, E. H., and TOOLE, V. K. (1952). A reversible photoreaction controlling seed germination. *Proc. natn Acad. Sci. U.S.A.*, 38, 662–6.

CANHAM, A. E. (1966). The fluorescent tube as a source of night-break light. *Expl. Hort.*, 16, 53–68.

CHAILAKHYAN, M. H. (1936). On the mechanism of photoperiodic reaction. *Dokl. Akad. Nauk U.R.S.S.*, 10, 89–93.

GARNER, W. W., and ALLARD, H. A. (1920). Effect of relative length of day and night and other factors of the environment on growth and reproduction in plants. *J. agric. Res.*, 18, 553–606.

HARDON, J. J., and DAVIES, M. D. (1969). Effects of vacuum-drying on the viability of oil palm pollen. *Expl. Agric.*, 5, 59–65.

HEIDE, O. M. (1970). Seed-stalk formation and flowering in cabbage I. Day-length, temperature and time relationship. *Meld. Norg. Landbrhisk*, 49 (27), 21.

MOLISCH, H. (1893). Zur physiologie des Pollens, mit besonderer Rucksicht auf die Chemotropischen bewegungen der Pollenschlaüche. *Sitzber. kais. Akad. Wiss. Wien. (Math – Nat. Kl.)*, 102, 423.

POST, K. (1953). It's a short night you want. *Bull., N.Y. St. Flower Grow.*, No. 99, 3–4.

SIEGELMAN, H. W., and FIRER, E. M. (1964). Purification of phytochrome from oat seedlings. *Biochemistry, N.Y.*, 3, 418–23.

Further Reading

EVANS, L. T. (1969). The Induction of Flowering – Some Case Histories. Macmillan, London.

VINCE-PRUE, D. (1975). Photoperiodism in Plants. McGraw-Hill, Maidenhead.

WAREING, P. F., and PHILLIPS, I. D. J. (1970). *The Control of Growth and Differentiation in Plants*. Pergamon Press, Oxford.

WILKINS, M. B. (Editor) (1969). *Physiology of Plant Growth and Development*. McGraw-Hill, London.

CHAPTER 6

THE FRUIT

The early stages of fruit development often depend critically on environmental factors and the crop will fail unless a good set is obtained. Inadequate pollination is a frequent cause of poor crop setting; low temperatures may prevent pollen-tube growth or cause insufficient insect activity to ensure pollen transfer. Fruit-setting compounds can however ensure the development of fruit in some species, notably tomatoes, even when low temperatures would preclude natural setting. Pollination is nevertheless the usual 'trigger' for the series of physiological processes that culminate in the formation of fruits.

Although fruits are classified by botanists into many different types according to their structure, these divisions are of little physiological significance. The layman's concept of a fruit as something that develops when the flower dies and that contains seeds is adequate for most physiological purposes.

Some fruits are of considerable commercial importance because they are good to eat or contain seeds used in cooking. Other fruits perform solely their prime function of fostering seed until it is able to provide the next generation.

FLOWER SENESCENCE AND FRUIT SET

The longevity of flowers, either on the plant or as cut blooms, is governed by several external and internal factors. Water supply and temperature can obviously play an important part, but as with leaf senescence (page 79) there is also some form of internal control. The events set in motion by pollination include an internally controlled acceleration of flower senescence. For example, pollination reduces the longevity of carnation (*Dianthus caryophyllus*) flowers, wilting of the petals occurring within 2 to 4 days of fertilisation. A study (Nichols, 1971) of flower senescence and fruit development in the carnation cultivar, White Sim, indicated that two substances (2-chloroethylphosphonic acid and 2,4-dichlorophenoxyacetic acid) accelerate the typical post-pollination behaviour of the flower, either by direct release of ethylene or by inducing its formation by the flower. Ethylene gas (0.2 vpm) produced the same effects.

The senescence of carnation flowers is delayed by treatments expected to inhibit either the production of ethylene or its physiological effects. Since the production of ethylene requires oxygen and the action of ethylene can be blocked by high concentrations of carbon dioxide, carnation flowers senesce more slowly in atmospheres of low oxygen (4 per cent) and high carbon dioxide (4 per cent) than they do in air.

Nichols also showed that the effects ascribed to ethylene do not depend on the presence of petals. This suggests that fruit does not develop as a result of petals dying and transferring their food resources to the fruit. It appears that there is a well co-ordinated series of events under a central control. Figure 6.1 illustrates the effect of chemical treatment on the mean dry weight of the ovary and on the time of petal wilting. It is clear that stimulation of ovary growth is accompanied by earlier petal wilting.

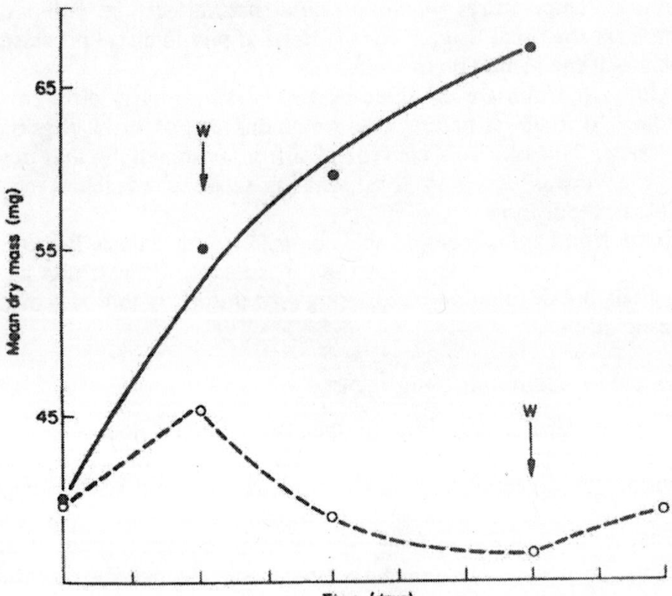

Figure 6.1 Increase in the dry weight of the ovary of carnation cv. White Sim; o control unpollinated; • flowers treated with 100 mg l^{-1} solution of 2-chloroethylphosphonic acid. W = petals wilting (from Nichols, 1971).

The flowers of a very wide range of species — from orchids to roses — produce ethylene but, more significantly, pollinated flowers of such diverse species as the lowbush blueberry (*Vaccinium angustifolium*) and the cultivated strawberry (*Fragaria χ ananassa* cv. Cavalier) produce four to five times more ethylene than their unpollinated flowers. The ethylene is produced in both these species by the stigma and style. Unpollinated flowers doubled their ethylene production when sprayed with a solution of auxin.

Pollen is a rich source of auxin and the growth of the pollen tube can stimulate the tissues of the gynaecium (the fruit) to produce further auxin. The use of synthetic auxins as 'setting' agents is yet another indication of the significance of the auxin group of hormones for fruit development. It would be premature however to suppose that the only role of the auxins here is in stimulating ethylene production.

Whereas auxins will promote fruit set in many species, particularly those of the Solanaceae and Cucurbitaceae, they will not improve the set of such horticulturally important subjects as apples, apricots, cherries, grapes, peaches or pears. These exceptions are sufficiently impressive to indicate that simple generalisations should be treated with caution. The unpollinated flowers of other species, hops (*Humulus lupulus*) and olive (*Olea europaea*), initially respond to auxin but fruit growth ceases after a short period. In these species a single application of auxin fails to trigger the normal sequence of events and growth continues only until the applied auxin is exhausted. In many of these species, however, sprays of gibberellins can ensure the development of unpollinated fruits, which are seedless. Certain seedless grape cultivars, such as 'Thompson's Seedling', 'Perlette' and 'Delaware', are usually treated commercially with gibberellins because of the resulting marked increase in fruit size. Developing seeds within fruits are rich sources of gibberellins. The set of apple, cherry and plum can be achieved by the use of a mixture of gibberellin (GA_3) and 2-(2,4,5-trichlorophenoxy)propionic acid (2,4,5TP).

There are many seedless cultivars of edible-fruit species in cultivation, such as seedless bananas, grapes, pineapples, cucumbers, oranges and other citrus fruits. Some of these fruits can develop without pollination, but what is more interesting is that no seed develops even if they are pollinated. It appears that in these seedless cultivars the hormone balance necessary for fruit development is largely achieved without the 'trigger' action of fertilisation. Seedless fruits produced by such cultivars or as a result of hormone treatments are termed **parthenocarpic**.

THE GROWTH OF THE FRUIT

The development of the fruit in many species begins when the flower is initiated and is quite advanced when the flower opens. The full form of the fruit is discernible below the petals of the female flowers of cucurbits. Cell division usually ceases as the flower opens and pollen is shed (anthesis), and further increase in the size of the fruit is mainly due to cell expansion. The cells of the water-melon (*Citrullus vulgaris*) become so large that they can easily be seen with the naked eye.

In spite of this changeover from increasing cell numbers to increasing cell size there is no discontinuity in growth, provided the stimulus usually given by pollination occurs. In fruits with few ovules, typically those of stone fruits such as plums, three phases of growth can be distinguished.

During the first phase the fruit enlarges rapidly and the integuments of the seed develop; the embryo remains small however. In the second phase the embryo develops but the overall size of the fruit does not increase. In the third phase the fruit increases in size again.

Although these development phases concern different tissues at different times and are real, their significance may have been somewhat exaggerated: they appear more marked if one of the favoured methods of measuring fruit growth — namely by measuring change in volume — is adopted. If increase in volume is accepted as a satisfactory measure of growth, the growth of a single fruit can be recorded without detaching it from the parent plant. This is attractive as it reduces sampling errors and can easily be done by immersing the fruit in water and measuring the amount of water displaced. However, during the second phase of fruit growth the locules, formerly occupied by low density integument tissues, are replaced by higher density endosperm or embryo tissues. Thus, although there is little change in fruit volume, fruit weight continues to increase at an almost unchanged rate. Figure 6.2 is based on some published data for the fruit of *Coffea arabica* (Cannell, 1971) and shows that the marked double sigmoid nature of the curve for fruit volume is only slightly indicated by the curve for increase in fruit weight.

At the end of the third phase the fruit matures and then senesces. At maturity many fruits fall from the parent plant because an abscission layer

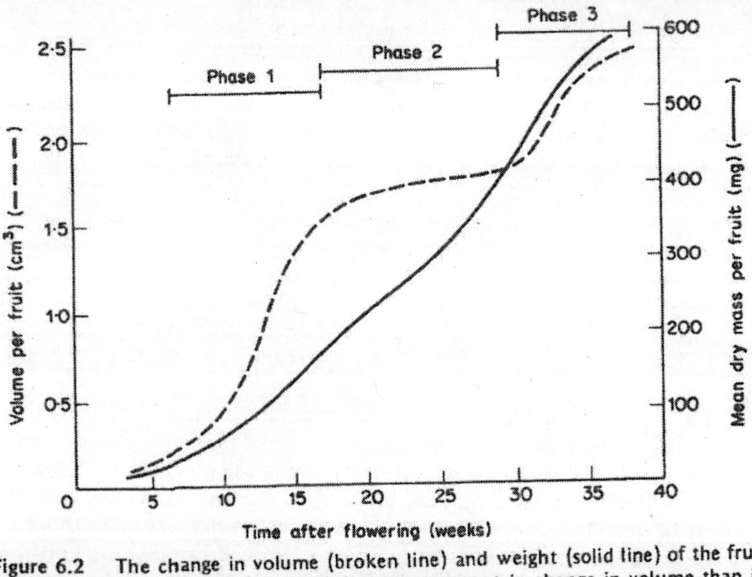

Figure 6.2 The change in volume (broken line) and weight (solid line) of the fruit of *Coffea arabica*. The phasic growth is more apparent in change in volume than in change in weight. See text (based on Cannell, 1971).

forms. However, before we leave the fruit rotting on the ground we should consider the source and the magnitude of the energy needed for fruit growth.

RESPIRATION

The oxidative breakdown of organic substrates, termed respiration, provides the bulk of the energy for fruit growth. Some fruits contain chlorophyll and can photosynthesise to contribute to their growth, but respiration of material translocated into the fruit generally provides most of the energy needed.

When respiring cells have access to air (as is normally the case), oxygen is consumed and carbon dioxide formed. We can summarise the overall respiration of a hexose as

$$C_6H_{12}O_6 + 6O_2 \rightarrow 6CO_2 + 6H_2O + 2817\ kJ$$

Many intermediate steps are involved at the cellular level and readers interested in this aspect of respiration should refer to more specialist books, such as that by Öpik (1980). We are more concerned here with the rate at which energy is being provided for growth and this can be assessed conveniently by measuring either the amount of oxygen absorbed or the amount of carbon dioxide evolved. If both are measured, then the ratio volume of carbon dioxide/volume of oxygen can be calculated and gives some indication of what substrate is being respired. The above equation gives a respiratory ratio or quotient of unity, which is typical of carbohydrate substrates. Fats have less oxygen per unit of carbon than carbohydrates, so they need more external oxygen to evolve a unit of carbon dioxide. Thus fats are characterised by respiratory quotients of less than unity, typically 0.7 to 0.8. On the other hand organic acids, such as malic acid which accumulates in many fruits, contain more oxygen than hydrocarbons and give respiratory quotients greater than unity (malic acid gives a value of 1.33).

The rate of respiration

Rates of respiration are expressed as volumes of gas per unit time. The volumes can be measured using an infra-red gas analyser, gas chromatography, and, for oxygen, paramagnetic analysers. Alternatively, the respiratory gases may be measured by gravimetric methods in which oxygen and carbon dioxide are selectively absorbed by chemicals and their resultant change in weight determined.

Temperature effects

For fruits and most other plant organs the respiration rate doubles — or slightly more than doubles — for every 10°C rise in temperature within the range 10 to 30°C. In other words the Q_{10} for the respiration in the range 10 to 30°C lies between 2.0 and 2.5. The rates achieved at any temperature within this range are usually steady, although a few hours may be needed to establish these steady states.

At temperatures above 30°C higher rates of respiration are initially established but the rate usually declines after a few hours to a level below that of the steady state at 30°C. This may be because either the tissues are not permeable enough to allow the diffusion of sufficient oxygen or the supply of substrate is limiting. These reasons are likely to apply to temperatures in the range 30 to 40°C, but at still higher temperatures inactivation of some of the enzyme systems occurs, leading to death at very high temperatures.

At temperatures below 10°C but above 0°C various effects may occur, depending on the particular species. With tropical or sub-tropical fruits there is some injury (termed **chilling injury**) manifest at these temperatures (see page 81). It seems probable that a contributory factor in this injury is a change in the pattern of respiration, such that harmful end products are formed at low temperatures but membrane changes are also implicated. Most of the changes involved in chilling injury are from a practical point of view irreversible. An exception is sweating of potatoes at low temperatures which can be reversed by a period at a higher temperature. These shifts in the respiratory pattern at low temperatures make it more difficult to generalise about the Q_{10} in the 0 to 10°C region. Nevertheless, bearing in mind that tropical and sub-tropical fruits are likely to be the exceptions, the Q_{10} for respiration is still about 2.0.

Thus, if we are to obtain meaningful measurements of the respiration rates during the development of fruits or other organs the measurements must be made at one or more constant temperatures. If only one temperature is used it should be within the temperature range normally encountered by that fruit during its development. The results of such a study can be expressed in many different ways. Consideration of figure 6.2 will make it apparent that we would see different trends in respiration rates if these were expressed either per unit of weight or per unit of volume.

Respiratory load

If the purpose of this type study is to determine the load placed on the parent plant at different stages of fruit development, then rates per individual fruit would be the most appropriate way of expressing the results. Cannell's (1971) study of the respiration and growth of the fruit of *Coffea arabica* had this intention. Coffee is a non-seasonal fruit crop when grown near the Equator (short days) in Kenya. Alternating wet and dry seasons nevertheless produce some phasing of cropping and biennial bearing is also a problem. Pruning and irrigation can reduce this unevenness of cropping, but Cannell considered that an understanding of the energetics of fruiting might lead to improved management. Figure 6.3 shows that respiration rate per unit fruit does not increase continuously

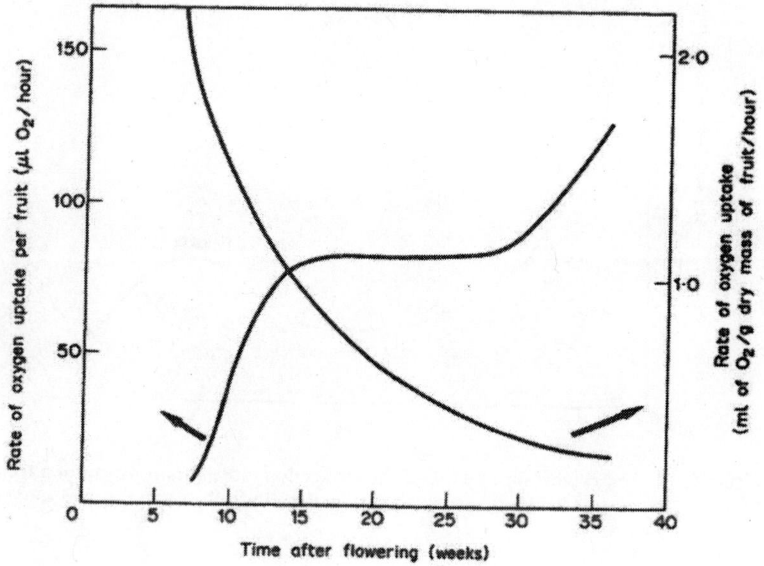

Figure 6.3 The rate of oxygen uptake by the developing fruit of *Coffea arabica*, expressed as per fruit and per gram of dry weight (based on Cannell, 1971).

as in many other species. The relatively steady rate from weeks 17 to 30 corresponds to the period when dry matter accumulates within the existing structure of the fruit. Cannell concluded that the rapid rise in respiration rate per unit fruit that occurs during the first six weeks after flowering indicated that the trees must be managed in such a way as to ensure large supplies of carbohydrates during this period. Fruit abscission was known to occur during this period if there were a carbohydrate shortage. The fruits respire progressively more rapidly from week 30 to maturity, and Cannell suggests that the carbohydrate drain on the tree could be reduced by harvesting the coffee fruits as soon as the pericarp becomes soft enough to be removed easily during processing.

Climacteric

Fruits generally decline similarly in respiration rate per unit weight as coffee in figure 6.3, but in some species the fruits show a marked increase in this rate just as they are ripening. This increase is called the climacteric (figure 6.4) and is followed by a period of decreasing rates of respiration, during which the senescence and breakdown of tissue is occurring. Ethylene has been shown to trigger the respiratory climacteric which coincides with a similar climacteric of ethylene production.

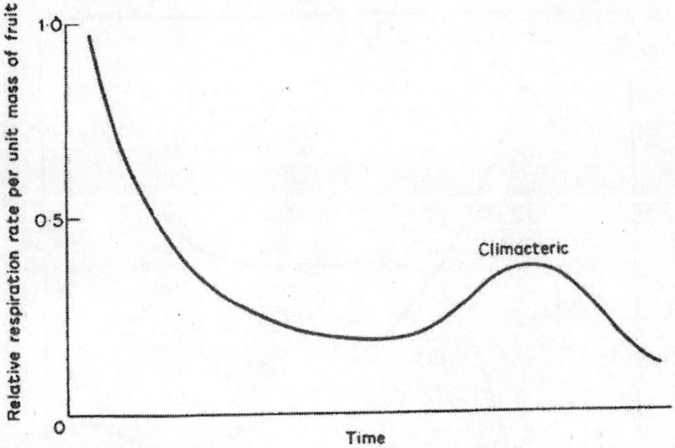

Figure 6.4 The typical time course of the respiration rate per unit weight of a fruit, showing the climacteric rise that occurs at maturation in fruit of some species.

Some fruits do not have a climacteric and their respiration rate continues to decline gradually; examples are grapefruits, oranges, lemons, grapes, pineapples and strawberries. Nevertheless many other fruit do show this climacteric rise, which has been extensively studied in apples and reported in fruits as diverse as those of the avocado and the tomato.

A study of the climacteric in tomato fruits (Chalmers and Rowan, 1971) indicates that the rise in respiration rate commences when the fruit has reached a mature size and is just beginning to change colour. The climacteric reaches its peak as the fruit reaches table ripeness. Chalmers and Rowan (1971) suggest that the climacteric is caused by a change in the membrane which surrounds the vacuoles and allows phosphate compounds, hitherto imprisoned and inactive within the vacuole, to become active. In their experiments, Chalmers and Rowan grew plants at different levels of phosphate nutrition and demonstrated that a low phosphate status produced a low climacteric rise in respiration. They suggest that this vacuolar mediation of phosphate availability may be a 'ubiquitous method of respiratory control in storage tissues'.

Changes in cell permeability have been widely reported to accompany the climacteric. Furthermore, the role of ethylene in promoting ripening of such fruits as banana and citrus, which do not show a climacteric, is also thought to involve a decrease in "organisation resistance" which is characteristic of ripening.

The best time to harvest apples is just before the climacteric rise in respiration. This time is characterised by the commencement of the conversion to sugar of starch, accumulated in the fruit during the summer.

A sequence of tests in which iodine solution is applied to the cut surface of fruits can help to identify the commencement and built-up of starch (North, 1971). The blue-black stain will first reveal starch around the core and under the skin but it will eventually fill the whole fruit. It was originally thought that this test would be particularly valuable for pears because their storage performance is much enhanced if they are harvested at optimum maturity. However, tests over several years are tending to show that it is no more reliable than adhering to a calendar date derived from experience.

THE STORAGE OF FRUIT

Lowering the temperature preserves stored fruit by reducing their respiration rate. Temperatures which would preclude freezing injury but would be as near 0°C as possible would at first sight seem best for prolonged life in store. Many fruits exhibit chilling injury however at temperatures below 10 to 12.5°C.

Cucumbers have a storage life of only 15 days at 5°C, but at the optimum temperature of 12.5°C they do not deteriorate until after 65 days. Their storage life at 20°C is about 50 days. Cultivars differed in their storage life, comparable figures being Ohio MR200, 10 days; Cubit, 20 days; and Marketer, 47 days. Ethylene production is stimulated at chilling temperatures but its relationship to injury or to cultivar sensitivity is at present unknown.

Apeland (1966) tried to 'condition' cucumbers in an attempt to enable temperatures below 12.5°C to be used to prolong their storage life. Holding the fruits at 12.5°C or 20°C for 1 to 4 days increased storage life at 12.5°C following a period of 8 days at 5°C. Less chilling injury was paralleled by a reduction in the amount of ethylene produced while at 5°C (figure 6.5).

Cultivars of apple differ in their susceptibility to low temperature injury and in their optimum temperature for storage. The fruit of the cultivars Blenheim Orange, Ellison's Orange, Idared and King Pippin should all be stored at 3.5 to 4.5°C, whereas the optimum temperature for fruit of Golden Delicious, Grenadier and Red Delicious is about 0°C. Potassium deficiency increases the susceptibility of apples and plums to low temperature injury, even if they are of a cultivar which can normally be stored at about 0°C.

Almost any treatment affecting the growth of apple fruit on the tree is likely to affect the storage life. As mentioned previously, differences in the storage behaviour of apples picked from the same tree at the same time can be related to the apples' positions within the canopy. The mineral composition of apple fruit, which is affected by growing conditions, is also known to have a considerable effect on storage performance.

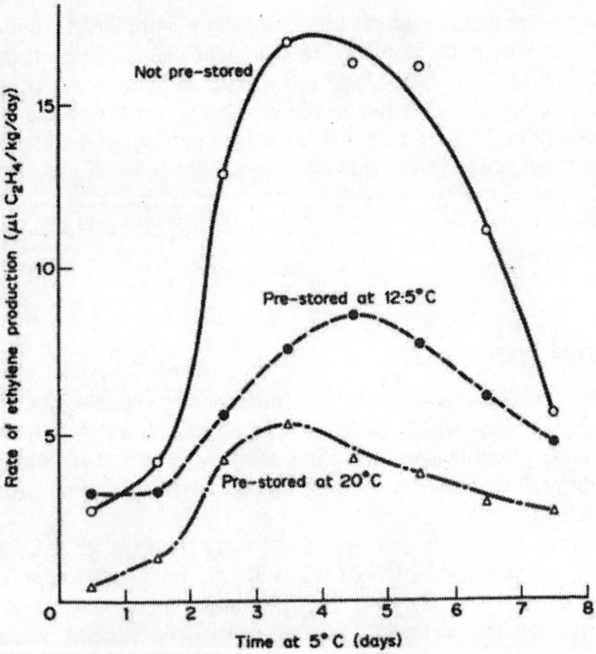

Figure 6.5 Ethylene production of cucumbers during chilling at 5°C with and without pre-storage at 12.5 and 20°C (from Apeland, 1966).

Gas or controlled atmosphere storage of apples

Kidd and West demonstrated as long ago as 1919 that apples stored at low temperatures in gas mixtures of 10 per cent carbon dioxide, 10 per cent oxygen and 80 per cent nitrogen had a longer storage life than similar apples stored in air. They termed this method "gas storage", but more recently the American term "controlled atmosphere storage" – often abbreviated to "CA storage" – has become widely used.

The principle involved seems attractively simple. If the tissues receive less oxygen, they will respire at a lower rate. Furthermore, if the carbon dioxide concentration is higher than normal, this will also tend to reduce the respiration rate.

Nothing is as simple as it seems, for the benefits CA storage cannot be explained solely in terms of reduced respiration rates. Even so it is reasonable to ask 'what would happen if the fruit were stored in an atmosphere of nitrogen only?'. Unfortunately, we have to contend with what is known as the **'Pasteur effect'**, which is illustrated in figure 6.6. Although Pasteur's observation was made more than a century ago, our understanding of the underlying biochemistry is still incomplete. The effect itself is quite straightforward: as the oxygen concentration falls, there is also a fall in the respiration rate (as measured by the volume of carbon dioxide released) until a critical low level of oxygen is reached (in figure 6.6 this critical level is shown as about 3 per cent). At lower concentrations carbon dioxide production increases greatly and can reach levels greater than that achieved in air. Therefore, the fruits would be quickly respired away if oxygen concentrations below about 3 per cent were used during storage.

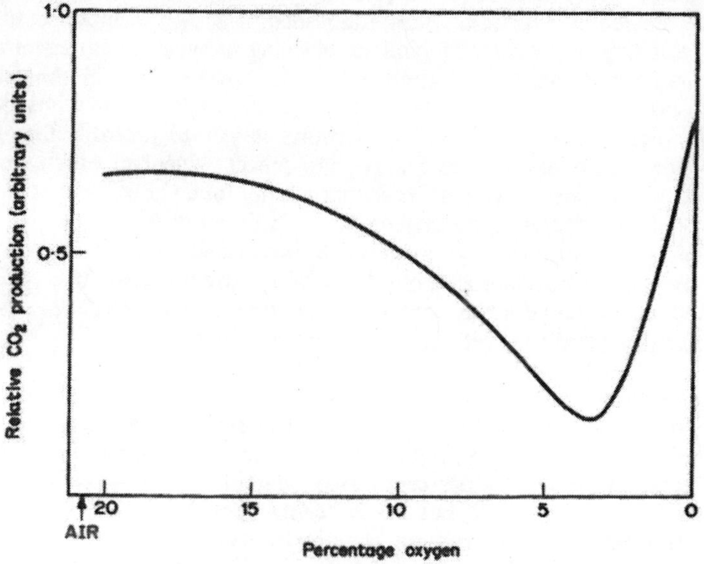

Figure 6.6 The Pasteur effect: as the concentration of oxygen is decreased, respiration also decreases until anaerobic respiration takes over and causes the rate to rise again.

It is also clear that respiration will change from a type requiring air (aerobic) to one not requiring air (anaerobic). Anaerobic respiration is also termed fermentive respiration, the end products of which are often alcohols. The fermentation of fruit to produce alcohols is of major economic importance, the greater part of the world production of grapes being used for this purpose.

Unfortunately it is not the concentration of oxygen in the atmosphere surrounding the fruit that is critical but rather the oxygen concentration *inside* the fruit at the sites where respiration occurs. This internal concentration is controlled by the external concentration, but differences between the fruit of various apple cultivars require the use of particular concentrations of oxygen. Higher than normal carbon dioxide concentrations can cause a condition known as 'brown heart' in pears and 'core flush' in apples. The storage life of some cultivars is not bettered by any artificial mixture of gases, so air is often the preferred storage gas.

Other cultivars of apple, notably Cox's Orange Pippin, have a greatly extended storage life if atmospheres containing less than 1 per cent carbon dioxide and 2.0 per cent oxygen are used at a temperature of 3.0 to 3.5°C (see Sharples and Stow, 1978).

Stores designed for controlled atmosphere storage are naturally more expensive than conventional cold stores, because they have to be gas tight. Once the required atmosphere is established, it is then maintained either by scrubbing out unwanted gases or bleeding in small quantities of any required gas. High concentrations of ethylene were found in controlled atmosphere stores of apples. Because of the known properties of ethylene in promoting the ripening of some fruits it seemed probable that this ethylene would have to be removed to prevent ripening, which would defeat the purpose of storage. It was eventually found however that at the storage temperatures used ethylene does not promote the ripening of the apples but recently it has been shown to increase scald, a skin disorder, in Bramley apples. Ethylene is so physiologically active that particular caution should be exercised when contemplating storage of different species of fruit or storage of fruit with other vegetable products.

In passing, it is important to realise that the principles of controlled atmosphere storage can be applied to the storage of any respiring organ. The storage life of cauliflowers and cabbages can be extended in this way. Isenberg and Sayles (1969) showed that atmospheres at 0°C and containing more than 5 per cent carbon dioxide and 5 per cent oxygen were little better than air, but 2.5 per cent of each gas prolonged storage life. The cabbages produced are abnormally sweet to the taste however. A more normal flavour, similar to that of freshly harvested field cabbage, is obtained with 5 per cent carbon dioxide and 2.5 per cent oxygen. Air-stored cabbages are bland and flavourless. The cultivars used were all of the Danish type, but despite this there were differences in storage life both between and within cultivars. These authors suggested that selection for suitability for CA storage would be worthwhile. They also stressed that successful storage depends on having a first class product — free from disease and damage — to put into store. This need for an initially high quality is a major general requirement for storage.

RIPENING AND ABSCISSION

One characteristic of many fruit is that they change colour as they ripen, which fortunately makes it easy to identify fruit that is ripe for harvesting. Colour change is a better guide to ripeness in those species whose fruit *develop* a pigment as the fruit matures, as opposed to those that lose a pigment. The loss of a pigment, usually chlorophyll, is more indicative of over-maturity and senescence than of ripeness. Thus, if a green fruit turns yellow it may well have passed the stage when it should have been harvested; an example is the fruit of cucumber. As examples of positive colour development in ripening fruit, we will take the tomato and the apple.

Tomato

Although cultivars of tomato have been bred with fruit ranging from yellow to red, in most commercial cultivars the ripe fruit is orange-red. The red pigment of the flesh is lycopene and superimposed is a yellow pigment in the translucent skin, together giving the fruit its orange-red appearance. Young fruit picked from the plant and placed at high temperatures fail to develop the red pigment in the flesh and so appear light yellow when 'ripe'. More mature (but still green) fruit will develop a red flesh colour at high temperatures, but the yellow pigment of the skin will not develop in the dark, causing the fruit to appear pink.

Piringer and Heinze (1954) investigated the effect of light on the development of this yellow pigment. Small portions of the skin can be removed from fruit and leached with acetone or petroleum ether to remove the carotenoids in the adhering cells; the remaining yellow pigment is clearly visible. With fruit which have been ripened in the dark, the skin prepared in this way is colourless. Piringer and Heinze (1954) found that only brief daily illumination with light of very low energies was needed to ensure the development of the yellow pigment, which was later identified as a flavonoid. Spectroscopy indicated that the formation of the pigment involves phytochrome (page 109), for the enhancement of colour by red light can be reversed by far-red light. The production of the yellow pigment is confined to the general area of light treatment; if a patch of skin is treated, then only that region develops colour.

Apples

The red colour of apple fruit also requires light for its formation. Although all parts of the spectrum cause some development of the red pigment (anthocyanin), the optimum wavelength lies in the 600 to 700 nm region. An initial period of about 20 hours' irradiation is needed before any pigment is formed, but thereafter the amount of pigment accumulated is directly proportional to the time at any constant level of irradiation.

Abscission

In addition to colour changes, the ripening of fruit is often accompanied by the formation of an abscission layer. Continuing to take apples as an example because their commercial importance means that we know quite a lot about them, we find that there are three major periods during fruiting when abscission can occur. These are usually descriptively designated 'post-blossom drop', 'June drop' (drop of young fruit), and 'pre-harvest drop'. There is no doubt that auxins play a major but inconsistent part in controlling this abscission of fruits: they promote 'June drop' and inhibit 'pre-harvest drop'. Nitsch (1953) has suggested that this seasonal difference in the action of auxin can be partially explained by differences in the physiological stage of the pedicel. 'June drop' abscission results from active cell division in the abscission layer, whereas 'pre-harvest drop' abscission appears to be due to a loosening of the cells. Nitsch considers it more significant that 'June' applications of auxin causes seed abortion in the apple (and pear), whereas the seed are sufficiently developed at a later time to be insensitive to applied auxin.

Once apple seed are fully developed, their role in supplying hormones is greatly reduced, allowing the abscission layer to form and causing 'pre-harvest drop'. Sprays of auxin at this stage top-up the diminishing supply from the seed and thereby delay abscission. The unripe seed is a source of auxins and gibberellins, which prevent abscission and mobilise materials for growth.

The precise role of ethylene in causing abscission or in promoting ripening is not understood, but its activity, particularly in conjunction with other hormones, is recognised in a wide range of species.

Seed within a fruit are not necessarily 'ripe' at the same time as is the fruit. Ripeness, particularly when applied to culinary fruit, is a state more significant to man than to the plant. Ripe for what? If for eating, then there is no reason for supposing that this stage in the continuous process of fruiting and dissemination of seed will have any actual physiological significance.

Even when the 'fruit' is intended for eating we may recognise more than one stage for harvesting. For example, pea pods (*Pisum sativum*) picked for eating fresh are much more immature than when harvesting is intended to produce dried peas for later consumption. In both cases, the pod is 'ripe' for harvesting. From the plant's point of view the fruit is not 'ripe' until its total function has been discharged; the seed would be less viable if harvested at an immature stage.

The embryos of some seed do not complete their development until they have been shed from the parent plant. The embyros in seed of the Orchidaceae consist of only a few undifferentiated cells when the seed is shed. In other species the seed are fully formed but require some period of 'after-ripening' before they will germinate. This and other aspects of the physiology of seed were discussed in chapter 1.

References

APELAND, J. (1966). Factors affecting the sensitivity of cucumbers to chilling temperatures. *Int. Inst. Refrig. Bull. Annexe*, 1, 325–33.

CANNELL, M. G. R. (1971). Changes in the respiration and growth rates of developing fruits of *Coffea arabica* L. *J. hort. Sci.*, 46, 263–72.

CHALMERS, D. J., and ROWAN, K. S. (1971). The climacteric in ripening tomato fruit. *Pl. Physiol., Lancaster*, 48, 235–40.

ISENBERG, F. M. and SAYLES, R. M. (1969). Modified atmosphere storage of Danish cabbage. *J. am. Soc. hort. Sci.*, **94**, 447–9.

KIDD, F., and WEST, C. (1927). Gas storage of fruit. *Spec. Rep. Fd Invest. Board (G.B.)*. No. 30.

NICHOLS, R. (1971). Induction of flower senescence and gynaecium development in the carnation (*Dianthus caryophyllus*) by ethylene and 2-chloroethylphosphonic acid. *J. hort. Sci.*, **46**, 323–32.

NITSCH, J. P. (1953). The physiology of fruit growth. *A Rev. Pl. Physiol.*, **4**, 199–236.

NORTH, C. J. (1971). The use of the starch-iodine staining test for assessing the picking date of pears. *Rep. E. Malling Res. Stn for 1970*, 149–51.

ÖPIK, H. (1980). *The Respiration of Higher Plants*. Edward Arnold, London.

PIRINGER, A. A., and HEINZE, P. H. (1954). Effect of light on the formation of a pigment in the tomato fruit cuticle. *Pl. Physiol., Lancaster*, **29**, 467–72.

SHARPLES, R. O., and STOW, J. R. (1978). Recommended conditions for the storage of apples and pears. *Rep. E. Malling Res. Stn for 1977*, 207–10.

Further Reading

LYONS, J. M. (1973). Chilling injury in plants. *Ann. Rev. Plant Physiol.*, **24**, 445–66.

PEREIRA, H. C. (Editor) (1975). *Climate and the Orchard*. Commonwealth Agricultural Bureaux, Farnham Royal.

WILLS, R. H. H., Lee, T. H. *et al*. (1981). *Postharvest: an Introduction to the Physiology and Handling of Fruit and Vegetables*. Granada, London.

INDEX